Fundamentals of Biometry

L. N. BALAAM, B.Sc., M.Sc.

Director of Biometric Services
Faculty of Agriculture, University of Sydney

Fundamentals of Biometry

A HALSTED PRESS BOOK

JOHN WILEY & SONS
New York

First published in 1972

Published in the U.S.A.
by Halsted Press, a Division
of John Wiley & Sons, Inc.
New York

Library of Congress Cataloging in Publication Data

Balaam, L N
  Fundamentals of biometry.

  (Science of biology series)
  1. Biometry. I. Title.
QH323.5.B35   574′.01′5195   72-4170
ISBN 0-470-04571-X

Printed in Great Britain
in 'Monophoto' Times Mathematics Series 569
by Page Bros (Norwich) Ltd
Norwich and London

Acknowledgements

Many of the examples and exercises in the text use real data from biological experiments and I wish to thank the various authors and scientific journals for permission to use their data.

I am indebted to the Literary Executor of the late Sir Ronald A. Fisher, F.R.S., to Dr. Frank Yates, F.R.S., and to Oliver & Boyd, Edinburgh, for permission to reprint Tables III and IV from their book *Statistical Tables for Biological, Agricultural and Medical Research*. I am also indebted to the *Biometrika* Trustees for permission to reprint the table giving percentage points of the F distribution.

I sincerely acknowledge the assistance given to me by present and past colleagues of the Biometry Section, Faculty of Agriculture all of whom have contributed in no small measure to this text. In particular I acknowledge Mr. D. G. McPherson, Lecturer in Statistics, University of Tasmania whose constructive criticisms during the writing of the manuscript I found most helpful. Mr. S. Abed, research assistant in biometry, assisted me with the answers to exercises which are found at the end of the text.

I am also indebted to Dr. F. H. C. Marriott, adviser to George Allen & Unwin, for his suggestions during the preparation of the final draft.

Finally, it is with a deep sense of gratitude that I acknowledge the late Mr. P. B. McGovern, Chief Biometrician, Queensland State Department of Agriculture, under whose patient tutelage for ten years I learned the fundamentals of biometry.

L. N. Balaam

Sydney, Australia,
July, 1970

Preface

In writing this book on biometry I have been motivated by the absence of a text on the subject which is suitable for an introductory undergraduate course. Most of the available textbooks are designed for graduate courses and present the analysis of variance as quickly as possible, since graduate students need this technique very early in their careers. This has not been the objective here. The book is intended to serve as an introduction to biometry and, as it has been written primarily for undergraduate students in the biological and agricultural sciences, it proceeds no further than the analysis of variance and simple linear regression. These topics are dealt with near the end of the text. This slow development of the subject is deliberate because I believe that, while many biological students have a psychological barrier to any subject related even indirectly to mathematics, this barrier disappears if biometry is developed slowly.

In this day and age at least an elementary course in biometry is essential in the training of all biologists. However, such a course cannot be presented without the use of the necessary tools. Thus, while no knowledge of the calculus is required for the understanding of my text, I have used the convenient Σ and 'dot notation' throughout. Also in accord with convention, letters from the Greek alphabet have been used for population parameters.

The exercises placed throughout and at the end of each chapter are intended to review the text material and to give practice in biometrical techniques. Also included at the end of each chapter is a section headed Collateral Reading. Here is to be found a list of textbooks and their chapters which contain material relevant to that presented in my text.

The volume has its origin in course notes for undergraduate and graduate students at the University of Sydney, modified in the light of experience over a number of years and also successfully used at Cornell University in the introductory sections of a graduate course. The underlying philosophy is aptly expressed by the following quotation contained in a personal communication from Professor H. A. MacDonald, Professor of Agronomy, Cornell University:

'*Statistical Analysis and Biological Research*

Statistical analysis is a mathematical technique used in the evaluation and interpretation of experimental data or findings. It should not be used just because it is the customary or prestigious thing to do. The investigator must understand its purpose, its place, its meaning, its use, its interpretation, and its limitations. Statistics is not biology; it is not the important research data or result, and is in no way the purpose of the study. The research result is never better than the investigator, the material or the techniques used. Statistics helps the investigator to evaluate his data, to guide his confidence in its use, and to improve his methods. Statistics can be the worthy servant of the biologist; it should not be his master.'

Contents

CONTENTS

–

Symbols and Notation

Symbol	Page first used	Explanation
z	85	Normal variable with mean zero, standard deviation 1
ν	90	Small Greek nu; degrees of freedom
$\hat{\mu}$, etc.	94	Hat; estimate of μ, etc.
θ	94	Small Greek theta
ε	95	Small Greek epsilon: true experimental error
H_0	98	Null hypothesis
H_1	98	Alternative hypothesis
ln	111	Logarithm to base e
∞	122	Infinity
β	134	Small Greek beta
D_j	134	Difference in paired observations
$\bar{d}.$	135	Difference in two sample means
τ	183	Small Greek tau
γ	183	Small Greek gamma
ρ	196	Small Greek rho; population correlation coefficient

CHAPTER 1

Some Basic Concepts

1.1 Introduction

It is important that at the beginning of a course of study in biometry or biological statistics that any preconceived ideas concerning this subject should be examined closely, since it is very likely that these ideas are incorrect. In the minds of many students, biometry has 'something to do with statistics'. This is true, but various dictionaries define statistics as 'a collection of facts, arranged and classified, respecting the condition of the people in a state; the science that has to do with the collection and classification of such facts'. Thus, frequently the term 'statistics' conjures up a vision of masses of figures and data, arranged in tables such as are found in the Year Books published by organizations such as the Bureau of Census and Statistics.

While it is important that data such as these be available, this text does not deal with this type of material which is called 'descriptive statistics'.

Biometry was defined by Treloar (1936) to be: 'the application of the statistical method, or mathematical logic, to the analysis and interpretation of biological variation'.

Fisher (1948) wrote (a reprint of this paper is contained in *Biometrics*, (1964), **20**, 2—an issue of *Biometrics* which is devoted to the outstanding contribution of Fisher to modern biometric science):

'The rise of biometry in this twentieth century, like that of geometry in the third century before Christ, seems to mark out one of the great ages or critical periods in the advance of the human understanding. From its humble beginnings in meeting the needs and satisfying the practical requirements of the work of land measurement and architecture, geometry found its way, by the enchanting clarity of its concepts and processes, into the heart of what the Greek world meant by a liberal education; an education that is fit for free men who would think for themselves, and not fit only for slaves and

1

officials whose aims and concepts were dictated from above. It was the liberation of the spirit experienced by the Greek students of geometry which gave the subject to them the exalted status it undoubtedly held, and won the veneration of the entire period. ...

'Men learnt to reason, deductively, from well defined abstract concepts, to cogent and irrefrageable conclusions. And with its use, with its exercise, in the field of geometry, the principles of deductive reasoning came to be understood, or at least to be codified, so as to give rise to the subject known as Logic. ...

'Now I suppose circumstances might have conspired to give to surveying, or to astronomy, or to any other subject sufficiently rich in observational detail, the honour of compassing the second great stage of intellectual liberation by making known the principles of that second and scarcely explored mode of logic, which we know as induction; of clarifying the principles of reasoning from the particular to the general, from the obervations to the hypothesis, in ways necessarily inaccessible to purely deductive logic, or to any mathematics which can properly be regarded as derivable wholly from deductive logic, or of making men free to recognize with certainty the consequence not of axioms or dogmas, but of carefully ascertained facts. But, as it happened, it has been reserved for biometry, the active pursuit of biological knowledge by quantitative methods, to take this great step'.

Fisher goes on to point out that it is the task of biometry 'to understand, design, and execute the forms of observation, surveys or experiments' and that 'the observational material requires interpretation and analysis'.

I define biometry as 'the development and application of statistical theory and methodology, or mathematical logic, to the design, analysis and interpretation of agricultural and biological experimentation'.

The adjective 'mathematical' which has been used in the definition may lead some to conclude rashly that this is a difficult and complex subject. However, this point of view could not be further from the truth. While advanced statistical theory requires advanced mathematical techniques, the general logical principles and the elementary statistical methods can be understood by any one with a knowledge of high school algebra provided he is prepared to spend some time studying same.

Because it is possible that an opinion different from this might be held, I would like to point out that the training of an agricultural scientist or research worker in one of the branches of the biological sciences is incomplete for at least two reasons, unless some attention is given to biometry. These two reasons are:

Firstly, the study of biometry will help the potential biologist or agricultural scientist to understand the nature of variability. The need for statistical method and an understanding of biometrical theory arises, in part, from the variability of the data which have to be handled. It can be claimed without fear of contradiction that, of the various characteristics of biological material, the one common to all is that of variability.

Secondly, biometry will teach how to derive general laws from small samples. It is because the biologist has generally to deal with small samples (the yields of a wheat strain grown on six or eight plots; the weight gains of four or five pigs fed on a particular diet) that a large section of modern statistics has developed. The process of inductive inference—reasoning from the particular to the general —has had considerable influence in statistical science. Prior to the work of 'Student' (1908) who developed what is known as the 't' distribution, there existed a great gap between the biological research worker and the mathematician. Professor Sir Ronald Fisher was mainly responsible for bridging this gap and any student in Biometry would be well advised to read his texts—'Statistical Methods for Research Workers' and 'The Design of Experiments'.

Goulden (1952), writing on the principles of experimental design, says of Fisher's type of statistical reasoning:

'There was a complete recognition of inductive logic and at the same time a presentation of statistical tools that could be applied without breaking faith with the mathematical principles of probability. Only history will give the complete picture, but a personal opinion can be ventured that this point of view, presented by Professor Fisher and since developed by many others, was the primary cause of a change of attitude on the part of experimentalists towards statistical methods, and the rapid increase in the application of statistical methods in practically all research laboratories'.

The increase in the application of statistical methods and statistical theory has not taken place only in the biological sciences, as evidenced by the development of such subjects as econometrics and psychometrics. Rowe (1960) writes: 'It is of course true that without judgment and common sense statistics can be made to mean anything, but to remove statistical methods from medicine, science, economics and other fields of learning would be to lose a powerful weapon in the pursuit of understanding.'

Huff (1954), in an interesting and amusing text, gives many examples of how the misapplication of statistics can result in erroneous conclusions. These examples serve to illustrate the dangers inherent in the 'cookery-book' approach to the subject of statistics. This type of approach can be particularly misleading where the design and analysis of experiments is concerned.

Attempts to learn one or two methods of analysis, or to search through a statistical-methods textbook to find what is thought to be an appropriate method of analysis, will lead inevitably to innumerable errors. It is probably correct to say that statistics is often the most misused and abused science of the present-day. For this reason, a real understanding of the principles underlying the various methods is essential if errors are to be avoided. Nowadays, all research organizations worthy to be called such have consultant statisticians, and if the only knowledge of this subject is that obtained here, very good advice would be that all except the most elementary problems should be taken to a statistician or biometrician. On the grounds of courtesy, this should be done before an investigation is started. If this is not considered a sufficient reason then selfishness might be a motive, because no statistician can give the best possible service if asked for advice only when an experiment or survey has ended and all the data have been collected. This is so since the design frequently determines the method of analysis.

While the reader of this text will not hope to be a professional biometrician, he will at least have an appreciation of a number of statistical techniques which are used widely in research. These techniques will be some of the tools used in his own research and will not be the research itself. The research might be in agronomy, animal husbandry, genetics or plant breeding, etc. It is with this in mind that the text has been written.

1.2 Two Simple Experiments

In the simplest situation, the aim or purpose of an experiment is to compare the effects of two treatments. Generally, more than two treatments are studied simultaneously. The word 'treatment' is used to indicate any kind of factor whose different states are under investigation, e.g. the 'treatments' might be levels and/or types of fertilizer (or insecticides), methods of cultivation or pruning, planes of nutrition, breeds of animals or strains of a cereal.

Consider a simple experiment designed to test the efficiency of a new insecticide proposed to kill mosquitoes. The efficiency of the new treatment (insecticide) will be determined by treating a group of mosquitoes and recording the number (or proportion) of insects that die in a subsequent period of time. This number (or proportion) would be compared with the number (or proportion) of mosquitoes which die in a similar group treated with the present standard insecticide (this latter group is usually called the *control* group).

Suppose that 200 insects are available for the experiment and suppose that 100 insects are allotted 'at random' to each of the treated and control groups. Each insect, by the draw of a card or the use of a table of random numbers, is given an equal chance of being in either group, subject to the restriction of there being 100 insects in each group. If the 200 insects were numbered, this could be achieved by using cards numbered 1 to 200, shuffling and drawing 100 cards. These 100 insects would then form, say, the control group, the remainder the group to which the new insecticide will be applied.

'Random allocation' or *randomization* is one of the requirements for a correct experimental design and is a common-sense safeguard. It has been said that randomization is somewhat analogous to insurance in that it is a precaution against disturbances which may or may not occur and which, if they do occur, may or may not be serious. Further, it must be emphasized that randomization is necessary if the results of mathematical and statistical theory are to be applied to the observed results by the methods outlined in this text.

It would, of course, be foolish to take two groups of insects from different sources, or to use two different strains and to have one group from one source or one strain as the treated group and the other as the control, since then it would not be possible to state if any observed difference in the proportions for the two groups were due to either the treatment on the one hand, or to the source or strain on the other. The statistician would say the effect of treatment was confounded with the effect of source or strain.

To return to the simple experiment under consideration, suppose that the results of the experiment are as shown in Table 1.1.

TABLE 1.1   *Results of hypothetical experiment on efficiencies of insecticides*

| | \multicolumn{3}{c}{Number of insects} | | |
Group	Dead	Surviving	Total	% Dead
Treated	85	15	100	85%
Control	75	25	100	75%
Total	160	40	200	80%

Examination of Table 1.1 gives the impression that the new insecticide is superior to the standard since the difference between the percentages dead in the treated and control groups is 10%. But can the experimenter be reasonably certain that this 10% is not due to chance, or, as is often said, due to random variation? How often, if the new insecticide were not superior to the standard and if the experiment were repeated, would a difference as large as (or larger than) that observed be obtained?

The answer to these and similar questions will be found in the study of biometry.

However, at this introductory stage consider yet another simple experiment. Suppose that it is desired to test a new fertilizer against a standard fertilizer for a particular cereal, and that the basis for comparison is the grain yield. If an area of one acre is available for the test, the simplest procedure would be to divide the area into two equal parts, to apply the new fertilizer (treatment A) to one half and the standard fertilizer (control) to the other, and to obtain the yield from each half separately. Suppose the yield from A is 1·25 bushels (i.e. 2·5 bushels/acre) greater than that from the control. This could indicate that A is superior to the control, but how reliable is this evidence?

It will be appreciated that no matter how carefully a *uniformity trial* (i.e. a trial in which the area under consideration is divided into a large number of plots and all plots receive the same treatment) is set out, the yields from equal areas in different parts of the field will vary because soil heterogeneity is always present. Table 1.2 gives the yields from a typical small uniformity trial.

TABLE 1.2 *Uniformity trial data*

Yield in bushels/acre of 20 plots of barley

35·0	35·5	36·2	37·3
34·8	35·0	33·5	35·1
33·0	31·2	32·5	33·1
31·4	30·5	30·5	25·4
29·6	28·4	29·5	30·4

Since in the case of a uniformity trial the yields vary from plot to plot, would the observed difference of $2\frac{1}{2}$ bushels/acre in the hypothetical experiment indicate a real difference in the yielding ability of the two treatments? For the results of such a trial to be absolutely conclusive and for the experimenter to be 100% certain, he would have to be sure that if the two half-acre plots had received the same treatment there would have been no difference in yield.

If the experiment is to be carried out and the two treatments tested, it will not be possible to have the same treatment on each of the two half-acre plots. Thus, there is no way of being 100% certain that the observed difference does indicate a superiority for treatment *A*. Further, results of this simple trial are rendered useless in the sense that they are not suitable for statistical analysis—there is no way of assessing the probability that a difference as large as (or larger than) that observed could be obtained if the treatment were not superior to the control.

The reliability of any yield figures and hence of any yield differences depends on a great many factors, each of which contributes in a smaller or larger way to the determination of the yield. Some of these factors are under the control of the experimenter while others are not. Examples of the former are making sure that the plots (i.e. the units of tested material) are as nearly alike as possible, that the rates of sowing are the same (unless of course these are the treatments) and that the correct amounts of fertilizers are applied as uniformly as possible. Factors outside the control of the experimenter could include soil irregularities due to changes in type, undetected fertility gradients, drainage, insect damage or disease incidence. While these latter may be outside the control of the experimenter, they should at least be defined in properly designed and executed experiments.

Correctly designed experiments enable both an estimate of the variation due to uncontrolled factors to be made, and the results of mathematical theory to be used to assess the statistical significance of the observed treatment differences. (The practical significance of any difference is outside the province of mathematics.)

The first necessary modification of the simple procedure of dividing the one acre into two half-acre plots is to divide it into a number of smaller plots, say twenty plots each being 0·05 acre. Treatment A is allocated at random to ten of these plots and the control (C) to the remaining ten. The allocation of treatments to plots might be as shown in Table 1.3. Treatments A and C are now said to be replicated, there being ten replications of each treatment.

TABLE 1.3   *Randomization of field experiment*

A	C	C	A
C	A	C	A
A	C	A	C
C	A	A	C
C	A	C	A

Thus a second requirement for a correct design, namely *replication*, has now been introduced. It will be shown later how replication provides the basis in research for the estimation of random variability.

The yields from the twenty plots are recorded separately. Suppose now that the ten treatment A plots are averaged as are the ten control plots and suppose, again, the difference is 2·5 bushels/acre. Questions of the type asked on examination of Table 1.1 may again be posed. Can the experimenter be reasonably certain that this 2·5 bushels/acre is not due to chance or random variation? If treatment A and the control were not having different effects, how often would a difference between the means of two sets of ten plots as large as, or larger than, 2·5 bushels/acre be obtained due to chance?

As stated earlier, answers to these questions will be found as the study of biometry continues. However, the method of answering them can be simulated by the following consideration. First, assume that the treatment and control are not different and that data are available from a uniformity trial of twenty plots, treated with the control and carried out on the same site and under conditions identical with those of the experiment. Suppose that from the twenty plots a group of ten plots is chosen at random and that these plots are called 'treatment A' plots, the other ten plots the 'control' plots. The means of the two groups of ten plots are found and then the differences between the two means calculated by subtracting the 'control' mean from the 'treatment A' mean. If the mean yield of the ten 'treatment A' plots is different from the mean yield of the ten 'control' plots, this difference can only be due to random or unexplained variation since, as the data are from a uniformity trial, both groups of plots received the same treatment.

Suppose two new groups of ten plots are chosen at random to be called respectively the 'treatment A' plots and the 'control' plots, the means found for the two groups and a new difference between means calculated. Generally, this difference will be different from the previous one.

Suppose, further, that this process of randomly selecting 'treatment A' and 'control' groups and of obtaining the difference is repeated a very large number of times. The results of such a process could be tabulated as shown in Table 1.4.

Table 1.4 gives the number of times differences of various sizes were obtained when the above process was carried out 300 times. Examination shows that differences less than or equal to -2.50 occurred due to chance in 6 out of the 300 while differences greater than or equal to 2.50 occurred in 7 out of the 300. Thus due to chance differences less than or equal to -2.50 or greater than or equal to 2.50 occurred $(6 + 7)$ or 13 times in 300. Hence, as a result of random variation, differences less than -2.50 or greater than 2.50 could be expected to occur in approximately 5 % of trials.

Similarly, Table 1.4 shows that the mean of the 'treatment A' group exceeds the mean of the 'control' group by 2.00 in 5 out of every 100 trials (i.e. in $4 + 4 + 3 + 2 + 2 = 15$ trials out of the 300) or once in 20 times.

If the information contained in Table 1.4 were available in assessing the results of the hypothetical experiment where the difference was 2.5 bushels/acre, it would be possible to say that since only once in 20 times would random variation or 'error' alone produce a

TABLE 1.4 *Frequency distribution of differences between mean yields (bushels/acre) of 2 random groups of 10 plots*

Class range of differences (bushels/acre)			Frequency	
		< -2.995	1	} 6
-2.995	to	-2.745	2	
-2.745	to	-2.495	3	
-2.495	to	-2.245	4	
-2.245	to	-1.995	11	
-1.995	to	-1.745	12	
-1.745	to	-1.495	11	
-1.495	to	-1.245	12	
-1.245	to	-0.995	17	
-0.995	to	-0.745	17	
-0.745	to	-0.495	19	
-0.495	to	-0.245	20	
-0.245	to	0.005	23	
0.005	to	0.255	21	
0.255	to	0.505	19	
0.505	to	0.755	18	
0.755	to	1.005	17	
1.005	to	1.255	15	
1.255	to	1.505	16	
1.505	to	1.755	15	
1.755	to	2.005	12	
2.005	to	2.255	4	
2.255	to	2.505	4	
2.505	to	2.755	15 {	3 } 7
2.755	to	3.005	2	
		> 3.005	2	
		Total	300	

difference as large as, or larger than, 2·0 bushels/acre, the odds against the observed difference being due to random variations alone are greater than 19 to 1. In agriculture this degree of odds is usually regarded as sufficient, and the observed difference would be taken as indicative of a real difference due to treatment.

It must be realized that in accepting this form of argument there is a small risk (5 %) that the observed difference will be taken to indicate a real effect when, in fact, it is due to random variation alone. If it is desired to reduce the risk of this type of error, odds of 99 to 1 or a

1 % level of significance can be used. Of course, if a 1 % level is adopted the chance of detecting a real difference is reduced. In biological and agricultural research, 5% and 1% are the most commonly used levels of significance.

Even if the necessary data from an appropriate uniformity trial were available, it would be extremely time-consuming to carry out the above examination or test for each experiment. Fortunately, statistical theory supplies a solution. *Statistics*, from which can be estimated the amount of random variation to be expected in any proportion of cases, can be calculated. Tables similar to Table 1.4 can be prepared if desired, but generally the only information required is that if the treatment is having no effect, a difference as large as, or larger than, some quantity (which is calculated) could be expected due to chance or random variation in 5 trials out of 100.

Returning now to the results given in Table 1.1, the following is an outline of the statistical argument known as a *test of significance* which would be applied to this set of data.

First it is assumed that the treatment and control are equally effective in killing mosquitoes. If this assumption is true any difference between the observed percentages of dead insects for the two groups is due to chance, and the best estimate of the percentage of mosquitoes which would die as a result of spraying is 80%—the estimate obtained from the total number of insects.

Then statistical theory and analysis are used to answer the question —How often if the insecticides are equally efficient would a difference as larger as 10% (or larger) be observed in the two groups? As no statistical theory is at present known, the answer to this question may be obtained from the following simulation experiment.

It will be appreciated that if the assumption of equality in efficiencies is true, the observation of 85 dead in one group of 100 insects, and 75 dead in a second group, is the chance variation which would result if 200 insects were sprayed with an 80% effective insecticide, and these 200 insects subsequently divided at random into two equal groups.

Suppose then that 200 marbles are put in a bag and that 160 (or 80%) of these are white and 40 are red. The white marbles represent dead mosquitoes and the red marbles the surviving mosquitoes. Suppose, after a thorough mixing, 100 marbles are drawn at random and of these the number of white marbles noted. The 200 marbles have thus been divided at random into two equal groups. The group drawn out can be regarded as the group of insects treated with the new insecticide; those left in the urn as the control group. The diference in the percentage of dead in the two groups can be noted and, after this the 200 marbles can be thoroughly mixed and another draw made. In this way the probability of getting, due to chance, a difference of 10% or greater in favour of the treated group could be calculated.

With this information an argument, similar to that used in the case of the treated and untreated plots, could be completed. This process would again be time-consuming and tedious.

Exercise 1.1 Carry out the sampling experiment suggested in the previous two paragraphs and prepare 100 tables similar to that of Table 1.1. What proportion of experiments gives a difference of 10% in favour of the treated group?

1.3 Tests of Significance

The type of argument used in the previous section is known as a *test of significance*. The first step in a test of significance is the formulation of a null hypothesis which, often, states that the treatment has had no effect or that the treatments are having the same effect. On the basis of this assumption the probability of obtaining, due to chance, results as extreme as or more extreme than those observed is calculated. If this probability is small (less than 5% or 1%), the null hypothesis is rejected and the observed result is said to be *statistically significant*.

A test of significance consists of the following four steps:

(i) Formulate the null hypothesis (often that the treatments are not different in their effects; frequently that population parameters have particular values).

(ii) Choose an appropriate level of significance (α), (in the agricultural and biological sciences it is usual to use either the 0·05 or 0·01 level).

(iii) On the assumption that the null hypothesis is true and by using statistical theory, calculate the probability of the observed results (and those less favourable to the null hypothesis) being due to chance or random variation.

(iv) If the probability, p, found in (iii) is less than that in (ii), the null hypothesis is rejected. If not, judgment is suspended. If $p \leqslant 0.01$, the result is said to be *highly significant*; if $0.01 < p \leqslant 0.05$, it is said to be *significant*.

1.4 The Place of Biometry in Scientific Investigations

Schinckel and Moule (1962) outline the principles of scientific method, with particular reference to the role of hypotheses and experiments and give the common sequence of events in scientific investigations, as follows:

'1. The assembly of existing information on a problem, or the conduct of a series of observations designed to provide data; a detailed survey may be required.
2. The development of a hypothesis on the basis of this information and on a knowledge of biological principles which are, or might be involved.
3. The design, conduct and analysis of experiments to test the hypothesis.
4. Interpretation of the results obtained and the formulation of new hypotheses or theories which are to be the subject of further experiments.'

In the interpretation of most experiments, Schinckel and Moule state that at least one of the following three questions arises:

'1. How large are the differences between treatments?
2. What is the relative importance of the variables studied with respect to total variation?
3. How reliable are the treatment differences which have been observed?'

From these two quotations, it will be appreciated that biometry is necessary in the third and fourth stages of most scientific investigations.

REFERENCES

FISHER, R. A. (1948). 'Biometry', *Biometrics*, **4**, 217–219.
GOULDEN, C. H. (1952). *Methods of Statistical Analysis* (2nd edition). John Wiley, New York.

HUFF, D. (1954). *How to Lie with Statistics.* Victor Gollancz, London.

ROWE, A. P. (1960). *If the Gown Fits.* Melbourne University Press, Melbourne.

SCHINCKEL, P. G. and MOULE, G. R. (1962). 'Some Principles of Field Experiments with Sheep'. *Proc. Aust. Soc. Anim. Prod.,* **4**, 190–194.

STUDENT (1908). 'The Probable Error of a Mean', *Biometrika,* **6**, 1–25.

TRELOAR, A. E. (1936). *An Outline of Biometric Analysis.* Burgess Publishing Company, Minneapolis, Minn.

COLLATERAL READING

BAILEY, N. T. J. (1959). *Statistical Methods in Biology.* English Universities Press, London. Chapters 1 & 4.

FINNEY, D. J. (1964). *An Introduction to Statistical Science in Agriculture* (2nd edition). Oliver & Boyd, Edinburgh; John Wiley, New York. Chapters 1 & 2.

STEEL, R. G. D. and TORRIE, J. H. (1960). *Principles and Procedures of Statistics.* McGraw-Hill, New York. Chapter 1.

CHAPTER 2

Variables and Variation

2.1 Introduction

As indicated in the previous chapter, variability is the most common characteristic of the data which are handled by the agricultural or biological scientist. In fact the need for statistical method arises from this variability. It will be realized that

    (a) grain yields of adjoining plots,
    (b) green weight yields of different pasture plots,
    (c) plant heights,
    (d) girths of different animals,
    (e) numbers of plants having particular flower colours,
    (f) numbers of insects with particular eye colours,
    (g) numbers of diseased or healthy plants,
    (h) numbers of seedlings from the crossing of two selected parents falling into different categories

show a certain amount of variation. Characteristics which show variation of this sort are called *random variables* or *variates*.

The variables studied may be either *continuous* (a to d) or *discrete* (e to h). A continuous variable is one that theoretically can take any value in a given range which may be finite or infinite. The yield of a plot of wheat, for example, could be 104·5 g and if the yield were measured more accurately, it might be 104·4836 g. In theory any number of decimal places can be taken, and thus there is no logical reason why the yield of the plot cannot take any value within the range zero to some (maybe large) positive value. Thus grain weight and green weight yields are continuous variables. Again, plant height is a continuous variable since in a change from 20 cm to 25 cm, plant height must pass continuously through all possible values between 20 cm and 25 cm.

On the other hand, a discrete (or discontinuous) variable is one for which the possible values are not observed on a continuous scale. Thus the variables (e to h) are discrete. There would never be any suggestion of 15·5 plants having yellow flowers. The scale of measurement is in discrete units, often whole numbers.

Exercise 2.1 Classify the following variables as continuous or discrete: possible yields of oats from a given field, number of students in various agriculture classes, possible outcomes from rolling two dice, number of peaches on a tree, milk yields of Jersey cows in their first lactation.

2.2 Notation

In order to develop a general theory, it will be necessary to use symbols. Instead of writing variable each time, let X denote the variable being studied. Thus X may on one occasion be the grain yield per plot and, on another occasion, the number of diseased wheat plants per plot.

Further, let X_1 be the yield of the first plot, X_2 the yield of the second plot, X_3 the yield of the third plot and so on. In general let X_i (read as X sub i) denote the yield on the ith plot. Hence, if the plots in the uniformity trial in Table 1.2 are numbered along the rows starting with the top left hand corner, then

$$X_1 = 35 \cdot 0, \quad X_2 = 35 \cdot 5, \quad X_3 = 36 \cdot 2, \quad X_4 = 37 \cdot 3,$$

and, if no account of the position of the plot was taken, a general statement relating to the yields would be

'Let X_i = yield of grain from the ith plot, $i = 1, 2, \ldots, 20$.'

A sign of operation used continually in biometry is the sign of summation, usually written as $\sum$ which is the Greek capital letter *sigma*.

The sum $X_1 + X_2 + X_3 + X_4 + X_5$ is written

$$\sum_{i=1}^{5} X_i$$

and is read 'the sum of all the X's from 1 to 5' (or 'summation X sub i, i running from 1 to 5'). The subscript is usually a letter taken from the middle of the alphabet. The letters h, i, j, k, l are those most commonly used as subscripts. The first value which this variable subscript assumes is written below the summation sign and the last above it. These are called the *limits of summation*. When it is obvious what these limits are, they are often omitted.

Example 2.1 Write out in full the summations represented by

$$\sum_{i=1}^{6} X_i \;;\; \sum_{i=1}^{n} f_i X_i \;;\; \sum_{i=1}^{3} X_i^2 \;;\; \sum_{i=1}^{5} f_i X_i^2 \;;\; \sum_{j=3}^{5} (X_j - d) \;;\; \sum_{j=1}^{r} (X_j - a)^2.$$

$$\sum_{i=1}^{6} X_i = X_1 + X_2 + X_3 + X_4 + X_5 + X_6.$$

$$\sum_{i=1}^{n} f_i X_i = f_1 X_1 + f_2 X_2 + \ldots + f_{n-1} X_{n-1} + f_n X_n.$$

$$\sum_{i=1}^{3} X_i^2 = X_1^2 + X_2^2 + X_3^2.$$

$$\sum_{i=1}^{5} f_i X_i^2 = f_1 X_1^2 + f_2 X_2^2 + f_3 X_3^2 + f_4 X_4^2 + f_5 X_5^2.$$

$$\sum_{j=3}^{5} (X_j - d) = (X_3 - d) + (X_4 - d) + (X_5 - d).$$

$$\sum_{j=1}^{r} (X_j - a)^2 = (X_1 - a)^2 + (X_2 - a)^2 + \ldots + (X_{r-1} - a)^2 + (X_r - a)^2.$$

Example 2.2 Express each of the following sums with summation sign, indicating the limits of summation.

(a) $X_4 + X_5 + X_6 + X_7 + X_8$;
(b) $Y_1^2 + Y_2^2 + Y_3^2 + \ldots + Y_m^2$;
(c) $cU_1 + cU_2 + cU_3 + cU_4$;
(d) $(b + dV_8) + (b + dV_9) + (b + dV_{10})$;
(e) $(O_1 - E_1)^2/E_1 + (O_2 - E_2)^2/E_2 + (O_3 - E_3)^2/E_3$.

(a) $\displaystyle\sum_{i=4}^{8} X_i$ or $\displaystyle\sum_{j=4}^{8} X_j$ or $\displaystyle\sum_{k=4}^{8} X_k$;

C

(b) $\displaystyle\sum_{j=1}^{m} Y_j^2$;

(c) $\displaystyle\sum_{i=1}^{4} cU_i$;

(d) $\displaystyle\sum_{j=8}^{10} (b+dV_j)$;

(e) $\displaystyle\sum_{i=1}^{3} (O_i - E_i)^2/E_i$.

Exercises

2.2   Indicate clearly the summation represented by

$$\sum_{i=1}^{5} a_i ; \quad \sum_{j=1}^{3} (X_j - a); \quad \sum_{i=1}^{n} X_i^2 .$$

2.3   Write down in the above condensed form the following sums, indicating the limits of summation.

(a) $f_1 Y_1^2 + f_2 Y_2^2 + f_3 Y_3^2 + \ldots + f_n Y_n^2$;
(b) $X_1 Y_1 + X_2 Y_2 + X_3 Y_3 + \ldots + X_n Y_n$;
(c) $X_1 + Y_1 + Z_1 + X_2 + Y_2 + Z_2 + \ldots + X_m + Y_m + Z_m$.

2.4   Show that

(a) $\displaystyle\sum_{i=1}^{n} (X_i + Y_i) = \sum_{i=1}^{n} X_i + \sum_{i=1}^{n} Y_i$;

(b) $\displaystyle\sum_{j=1}^{m} aX_j = a \sum_{j=1}^{m} X_j$;

(c) $\displaystyle\sum_{i=1}^{n} (X_i + a) = \sum_{i=1}^{n} X_i + na.$

2.5   For the data in Table 1.2, let

$$Y_j = \text{yield of grain of } j\text{th plot}, \quad (j = 1, 2, \ldots, 20)$$

and let the plots be numbered

$$
\begin{array}{cccc}
1 & 2 & 3 & 4 \\
5 & 6 & 7 & 8 \\
9 & 10 & 11 & 12 \\
13 & 14 & 15 & 16 \\
17 & 18 & 19 & 20.
\end{array}
$$

Find

$$\sum_{j=1}^{10} Y_j \; ; \quad \sum_{j=1}^{20} Y_j \; ; \quad \left(\sum_{j=1}^{20} Y_j\right)^2 \; ; \quad \left(\sum_{j=1}^{20} Y_j\right)^2 \Big/ 20 \; ; \quad \sum_{j=1}^{20} Y_j^2.$$

2.3 Double Subscripts

When data are classified according to more than one criterion, the use of double or even triple subscripts becomes necessary. For example, suppose that in a nutrition experiment five pigs from each of four litters are chosen, and to each of the five pigs in a litter one of the five treatments is randomly allocated. The weight gains in a certain period of time for the twenty pigs in the experiment can be presented in the following way.

| Litter | \multicolumn{5}{c}{Treatment} |
	1	2	3	4	5
1	Y_{11}	Y_{12}	Y_{13}	Y_{14}	Y_{15}
2	Y_{21}	Y_{22}	Y_{23}	Y_{24}	Y_{25}
3	Y_{31}	Y_{32}	Y_{33}	Y_{34}	Y_{35}
4	Y_{41}	Y_{42}	Y_{43}	Y_{44}	Y_{45}

In this table, Y_{23} is the weight gain of the pig from litter 2 and which had treatment 3. In general the weight gain of the pig from the ith litter and which had treatment j is Y_{ij}. Here i may take one of the values $1, 2, 3, 4$ while j takes one of the values $1, 2, 3, 4, 5$.

The sum of the weight gains of the five pigs from litter 2 is represented by

$$\sum_{j=1}^{5} Y_{2j}, \qquad \text{i.e. } Y_{21} + Y_{22} + Y_{23} + Y_{24} + Y_{25},$$

while the sum of the weight gains of the four pigs which had treatment 3 is

$$\sum_{i=1}^{4} Y_{i3}, \qquad \text{i.e. } Y_{13} + Y_{23} + Y_{33} + Y_{43}.$$

The sum of the twenty weight gains is

$$\sum_{i=1}^{4} \sum_{j=1}^{5} Y_{ij}.$$

In this text, $\sum_{j} Y_{ij}$ (where the summation is with respect to j) will be written $Y_{i.}$ while $Y_{.j}$ will be written for $\sum_{i} Y_{ij}$. The grand total, $\sum_{i}\sum_{j} Y_{ij}$, will be written as $Y_{..}$.

Exercise 2.6   For the data in Table 1.2, let Y_{ij} ($i = 1, 2, 3, 4, 5$; $j = 1, 2, 3, 4$) be the yield of grain from the plot in the ith row and the jth column.
Find

$$\sum_{j=1}^{4} Y_{2j}; \quad \sum_{i=1}^{5} Y_{i3}; \quad \sum_{i=1}^{5} Y_{i.}^{2}.$$

2.4   Populations and Samples

Because there is variability in the material which is being studied, the problem of defining the characteristics of groups must be faced. One of the aims of statistics is to find the best methods of describing the characteristics of groups of individuals. In studying a set of data, it is first necessary to decide whether it is to be considered as all the possible data or as a sample from a larger set.

A *population* is defined as *the total set of actual or possible values of the variable*. A population may be finite or infinite, and the variable continuous or discrete. Consider the problem of determining the 'average' butter-fat percentage per Jersey cow in a clearly defined district on a particular morning. The population being studied is a finite population of butter-fat percentages and not a population of Jersey cows! The population is finite because there is a finite number of butter-fat measurements made, even though every cow in the district is recorded. Such a population must be distinguished from the infinite population of all actual and possible measurements of butter-fat percentages. It is against the background of such an infinite population that tests of significance and confidence-interval statements (both of which will be developed later) are made. Such infinite populations form the background of thinking in the development of biometry. Fisher (1948) wrote:

'The idea of infinite populations distributed in a frequency distribution in respect of one or more characters is fundamental to all statistical work.'

One of the principal objectives of statistics is to draw inferences with respect to populations by the study of groups of individuals forming parts of the populations.

A *sample* is any finite set of items drawn from a population. The purpose of drawing samples is to obtain information about the populations from which they are drawn. For this information to have real value, the samples must be drawn in such a way that the results obtained are unbiassed. This is ensured by drawing the sample at random, i.e. in making up the sample, each individual in the population has an equal chance of being included.

A *random sample* from a given population is a sample, chosen in such a manner that each possible sample has an 'equal chance' of being drawn.

Quantities which characterize populations are known as *parameters* while those which characterize samples are called *statistics*. A *parameter* is a fixed quantity, not subject to variation, whereas a *statistic* is a *variate*, since in general different samples from the same population give different values of the statistic. Generally a statistic is sought which 'best' estimates the corresponding population parameter. Conventionally, parameters are designated by letters from the Greek alphabet while statistics are represented by letters from the English alphabet.

2.5 Measures of Central Tendency

The first and most common measure of central tendency is the *arithmetic mean*. Consider a sample of size n of weights of lambs. If X_i ($i = 1, 2, \ldots, n$) is the weight of the ith lamb, then the *arithmetic or sample mean* is defined by

$$\bar{x} \ (\text{or } \bar{x}_.) = \sum_{i=1}^{n} X_i / n; \tag{2.1}$$

$\bar{x}$ is a statistic and is an estimate of the mean, μ, of the population from which the sample was drawn.

If the population is finite with N members, the *population mean* is defined by

$$\mu = \sum_{i=1}^{N} X_i/N. \tag{2.2}$$

While samples may be large or small, they will generally have fewer items than a finite population. For this reason, n has been used in the first definition and N in the second to indicate a smaller number of items in the first than in the second.

Two other measures of central tendency are occasionally used. These are the *median* and the *mode*.

The *median* is that value of the variate for which 50% of the observations, when arranged in order of magnitude, lie on each side. It is the value of the middle variate in an ordered arrangement of the variates according to magnitude or, in other words, it is that value of the variate which divides the total frequency in the whole range into two equal parts. If the total frequency is even, the median is the arithmetic mean of the two middle values.

Compared with the arithmetic mean, the median places less emphasis on the minimum or maximum values in the sample or population. For example, in a survey of scientist's salaries, the best measure of central tendency might be the median rather than the arithmetic mean, since this latter could be affected by the very high atypical salaries of one or two scientists. Alternatively, in such a survey the mode might be used.

The *mode* is that value of the variate which occurs most frequently.

Example 2.3 The numbers of tillers per plant in a sample of ten plants were

$$10, 9, 10, 11, 9, 12, 10, 10, 9, 12.$$

What is the arithmetic mean and mode of the number of tillers per plant?

$$\bar{x} = \sum X_i/10$$
$$= (10+9+\ldots+9+12)/10$$
$$= 10 \cdot 2.$$

In this sample 9 occurs three times, 10 four times, 11 once, and 12 twice. Thus 10 is the mode.

2.6 Measures of Spread

While the arithmetic mean is the most commonly used measure of central tendency, the *variance*, or its square root the *standard deviation*, is the most commonly used measure of spread in biological statistics.

Where statistics are used in the physical sciences, the *range* (difference between largest and smallest values of the variable) and the *mean deviation* (shortly to be defined) are used reasonably often. In fact, the range is now used considerably more in biological statistics than it was in the past.

It may be thought that an indication of spread could be obtained by summing deviations from the mean, i.e. by $\sum_{i=1}^{N} (X_i - \mu)$ or $\sum_{i=1}^{n} (X_i - \bar{x})$. However both these are zero since in each case the sum of the positive deviations cancel with the sum of the negative deviations. To avoid this cancelling, absolute deviations from the mean might be considered, and an infrequently used measure of dispersion is the arithmetic mean of the absolute deviations, $\sum |X_i - \mu|/N$ for a finite population of size N, or $\sum |X_i - \bar{x}|/n$ for a sample of size n.

This quantity is called the *mean deviation*. However it is not nearly as important as the *standard deviation* and since it does not lend itself readily to algebraic treatment, it is seldom used.

For a finite population of size N, the *variance* is denoted by σ^2 ($\sigma =$ the small Greek letter sigma and this is read as sigma squared) and is defined to be

$$\sigma^2 = \sum_{i=1}^{N} (X_i - \mu)^2/(N-1). \tag{2.3}$$

The positive square root of the population variance is called the *standard deviation*.

For a sample of size n, the *sample variance* is defined to be

$$s^2 = \sum_{i=1}^{n} (X_i - \bar{x})^2/(n-1) \tag{2.4}$$

and the positive square root of the sample variance is called the *sample standard deviation*.

In some of the older text books the variances are defined with divisors N and n. However, the divisor $(n-1)$ is adopted here because it can be shown that the value of s^2 averaged over a large number of samples is independent of n and approaches the population value σ^2. As defined here, s^2 is said to be an unbiased estimate of σ^2. This is considered in more detail in sections 6.8 and 6.9.

The variance and the standard deviation give an indication of the extent to which the values of X are scattered or dispersed. When the values of X cluster closely around the mean, the variance is small; when the deviations from the mean are large, the standard deviation is large.

The importance to be attached to the size of the standard deviation clearly depends on the values of the variable. Thus a standard deviation (s.d.) of one foot for a population of heights of trees is of less significance than an equal s.d. for a population of heights of wheat plants.

The ratio of the standard deviation to the population mean, or the sample standard deviation to the sample mean, expressed as a percentage is called the *coefficient of variation*. It is an absolute measure of dispersion in the sense that it is independent of the unit employed. Using this coefficient, a comparison of the variabilities of populations or samples having different means may be made.

For agricultural and biological data, coefficients of variation of the order of 10% to 15% are common. For very homogeneous material this figure may be reduced to 5%, while a coefficient of variation of 25% would indicate very considerable variability.

2.7 Method of Calculation

In calculating the variance it is not usual to find each deviation from the mean, square, and sum. Thus in finding the variance of a finite population it is not usual to find $X_1 - \mu$, then $(X_1 - \mu)^2$, and repeat the process for $X_2, X_3, \ldots, X_N$ to find $(X_1 - \mu)^2 + (X_2 - \mu)^2 + \ldots + (X_N - \mu)^2$. This procedure would take too long, and if the mean has not been calculated exactly considerable errors can arise.

Now,

$$\sum_{i=1}^{N} (X_i - \mu)^2 = \sum_{i=1}^{N} (X_i^2 - 2\mu X_i + \mu^2)$$

$$= \sum X_i^2 - 2\mu \sum X_i + \sum \mu^2$$

$$= \sum X_i^2 - 2N\mu^2 + N\mu^2, \qquad \text{where } \sum X_i = N\mu,$$

$$= \sum X_i^2 - N\mu^2$$

$$= \sum X_i^2 - \left(\sum X_i\right)^2 / N.$$

Thus for the population variance,

$$\sigma^2 = \left[\sum_{i=1}^{N} X_i^2 - \left(\sum_{i=1}^{N} X_i\right)^2 \Big/ N\right] \Big/ (N-1).$$

Similarly for the sample variance (when sample is of size n)

$$s^2 = \left[\sum_{i=1}^{n} X_i^2 - \left(\sum_{i=1}^{n} X_i\right)^2 \Big/ n\right] \Big/ (n-1).$$

These last two formulae are used in practice.

The terms $(\sum X)^2/N$ and $(\sum X)^2/n$ are commonly called *correction terms*.

2.8 Linear Transformation of Variable

Where no calculating machine is available, the arithmetic involved in the calculation of the parameters (μ, σ^2) or statistics $(\bar{x}, s^2)$ can frequently be reduced or simplified by a linear transformation of the variable (X) which is being studied. A new variable, U, is introduced which is related to X by the equation

$$X_i = a + cU_i.$$

If the constants a and c in this equation are chosen wisely, the variable U may take small integral values (positive or negative) which are easily manipulated. For example, the following set of X could be transformed by selecting $a = 65$ and $c = 1$, so that $U_i = (X_i - a)/c = (X_i - 65)/1 = X_i - 65$.

X_i	61	63	64	65	67	68	69
U_i	-4	-2	-1	0	2	3	4

Again, the set of X in Table 2.3 (p. 36) could be transformed by putting $a = 84{\cdot}5$ and $c = 10$ so that $U_i = (X_i - 84{\cdot}5)/10$.

X_i	44·5	54·5	64·5	74·5	84·5	94·5...
$X_i - a$	−40	−30	−20	−10	0	10 ...
U_i	− 4	− 3	− 2	− 1	0	1 ...

The method of choosing a and c is explained in Example 2.5.

If then the mean and variance of the variable U are calculated, it is not difficult to prove that

$$\text{Mean of } X \text{ variable} = a + c(\text{mean of } U \text{ variable})$$

$$\text{Variance of } X = c^2(\text{variance of } U).$$

It is obvious that the mean and variance of series of smaller integers are more easily calculated than the mean and variance of numbers of the order of sixties or seventies. For this reason this particular technique is used quite a lot, especially where no electric calculating machines are available.

2.9   Frequency Tables

For ease of calculations and presentation of the data, it is often desirable when the total frequency in a finite population or sample is large, to summarize the data in a frequency table. Suppose there are N observations of a variable X. If N is large, it is likely that the N values of X are not all different. Suppose there are only k different values, $(X_1, X_2 ..., X_k)$ and that these occur resp. $f_1, f_2, ..., f_k$ times. The numbers $f_i (1, 2, ..., k)$ are called the *frequencies* of the X_i.

The assemblage of X_i with their associated frequencies f_i is called a *frequency distribution*. A typical frequency distribution is presented in Table 2.1.

TABLE 2.1   *Frequency distribution for variable X*

Value of Variable	Frequency
X_1	f_1
X_2	f_2
.	.
.	.
X_i	f_i
.	.
.	.
X_k	f_k
Total Frequency	N

With discrete variables, the observed values of the variable are recorded in the first column of a frequency table. However, with continuous variables some grouping of the observed values into classes is usually necessary. In these cases the class intervals or class limits are recorded in the first column.

The decision on the class intervals to be used with a particular continuous variable is based on a number of considerations. If f values of the variable lie in the interval, $X - 0.5c$ to $X + 0.5c$, and if in the calculations to be made from the frequency table it is assumed that there are f values of the variable equal to X, then there will be differences between the computations made from the frequency table and those which would be made from the ungrouped data. These differences or errors are a function of the class range c; in general, the larger the class range the larger the errors. The class range is the difference between the upper and lower limits of any class. In deciding on the class interval to be used, the accuracy required in the com-. putations from the table, the range in the variable and the total frequency have all to be considered. As a general rule, the class range c should be between one-third and one-quarter of the standard deviation. For small samples of size 20, the number of classes might be 8 to 10 while for large samples or finite populations with a total frequency of 300 (say), 20 classes should be sufficient. After the class range is chosen the various class limits are determined and the observed values of the variable sorted into the various classes. The frequency table is made up finally by entering the frequencies opposite the corresponding classes.

The formulae for finding the mean and variance from a frequency table are

$$\text{Mean} = \sum_{i=1}^{k} f_i X_i / N,$$

$$\text{Variance} = \left[\sum_{i=1}^{k} f_i X_i^2 - \left(\sum_{i=1}^{k} f_i X_i \right)^2 \bigg/ N \right] \bigg/ (N-1),$$

where $N = \sum f_i$ = total frequency.

2.10 Apparent and True Class Intervals and Class Limits

In setting up a frequency table, there is no rigid rule for presenting the class interval. If a plant breeder measures the heights of 100 wheat plants and each height is measured to the nearest centimetre, the classes might be 'labelled' (in cm) 50–59, 60–69, 70–79, 80–89 etc. (Example 2.5). The plant whose height is recorded as 74 cm will be put into the class 'labelled' 70 to 79 cm. This is the *apparent* range for this class as all plants whose heights are between 69·5 cm and 79·5 cm will be in this class. This is so because those plants whose heights are between 69·5 cm and 70·5 cm will have their height recorded as 70, while plants with heights between 78·5 cm and 79·5 cm will have their height recorded as 79.

The interval 70 cm to 79 cm is called the *apparent* class interval while 69·5 cm to 79·5 cm is called the *true* class interval. The true class interval should be used in all calculations made from a frequency table. It should be noted that the *limits* of the true class interval are mid-way between two recorded measurements—the lower limit (69·5 cm) of the true class interval being midway between the upper limit (69) of the preceding apparent class interval (60–69) and the lower limit (70) of the apparent class interval under consideration (70–79).

The class range is the difference between the upper and lower limits of the true class interval.

2.11 Relative Frequency Tables

If the data in two frequency tables are to be compared and if the total frequencies in the two tables are very different, comparison can be made more readily if the relative frequencies of the various classes in the table are found.

The *relative frequency* for a particular class is the frequency for that class divided by the total frequency. The sum over all classes of the relative frequencies in a relative frequency table is unity.

If, in the illustration in the preceding section, the plant breeder after measuring the heights of 100 plants has a frequency of 24 for the class 80–89, the relative frequency of this class is 0·24. Here a relative frequency table would be useful if for one strain the plant breeder had measured 100 plants while for another strain he had measured 300 plants.

2.12 Graphical Representation of Frequency Tables

The commonly used graphs to represent frequency tables are the *histogram*, *frequency polygon* and *bar diagram*.

In the histogram, which is used for continuous variables, the true class intervals are represented on the horizontal axis, which need not begin at zero. Above these intervals, rectangles are erected whose areas are proportional to the frequencies of the various classes.

For a distribution whose class ranges are all the same, these rectangles are obtained by marking on the vertical axis (which must begin at zero) either the frequency per interval or the frequency per unit range. The frequency per interval gives a quick visual appreciation of the number in any particular class. (See Figure 2.2.)

When drawing a histogram for a frequency distribution with unequal class ranges, since the areas have to be proportional to the frequencies, the vertical scale must show the frequency per unit range (instead of the frequency per interval). The frequency per unit range is the frequency per interval divided by the range of that particular interval.

Returning to the illustration of the plant breeder and heights of wheat plants, suppose the lower and upper classes were 40–59 and 110–129 with frequencies of 7 and 9 respectively. Thus the true ranges for these classes would be 20. Suppose that all the other classes (e.g. 60–69) had a range of 10 and that the frequency for the class 60–69 was 14. The heights of the rectangles (frequencies per unit range) to be drawn above the true class intervals $39{\cdot}5$ to $59{\cdot}5$, $59{\cdot}5$ to $69{\cdot}5$, and $109{\cdot}5$ to $129{\cdot}5$ would be respectively $7/20 (=0{\cdot}35)$, $14/10 (=1{\cdot}40)$, and $9/20 (=0{\cdot}45)$. The areas of the three rectangles, obtained by multiplying the width by the height, will be

$$20 \times 0{\cdot}35 = 7, \quad 10 \times 1{\cdot}40 = 14, \quad 20 \times 0{\cdot}45 = 9.$$

These areas are equal to the frequencies of the three classes.

If a number of histograms are to be drawn for distributions which have different total frequencies, instead of using the frequency per interval or frequency per unit range, it is preferable for ease of comparison to use the *relative frequency per unit range*. A further advantage in using the latter is the fact that the sum of the areas of the rectangles is unity and thus constant from one histogram to the next.

Irrespective of the scale chosen for the vertical axis, it is important that the histogram be self-explanatory. Thus the scale on each axis should be marked and each axis clearly labelled.

In a *frequency polygon*, instead of a rectangle of the appropriate height being erected over each class interval, a dot is put at the same height over the midpoint of the class interval. These dots are then joined by straight lines to form the frequency polygon. Vertical lines can be drawn from the horizontal axis to the dots above the midpoints of the first and last class intervals. While frequency polygons are more quickly constructed than histograms, they do not give as accurate a picture as that given by histograms, since the areas above the various intervals are not exactly proportional to the frequencies.

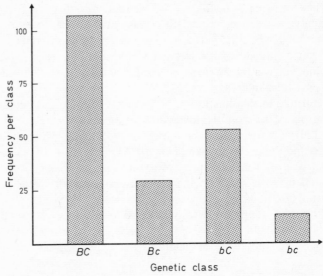

Fig. 2.1 Typical bar diagram

Another way to present data contained in a frequency distribution is by means of a *cumulative frequency curve* or *ogive*. The cumulative frequency to a particular value, X_q, of the variable is the sum of the frequencies corresponding to all values of the variable less than or equal to X_q, i.e.

$$\text{Cumulative frequency to } X_q = \sum f_i$$

where the summation is over those values of i such that $X_i \leqslant X_q$. The ogive is useful for determining the value of the median. In some instances, a *percentage ogive* is drawn. This is a curve of the cumulative relative frequencies expressed in percentages.

The *bar diagram* is often used with discrete or qualitative variables. It consists of equally spaced vertical rectangles of equal width, placed on a common horizontal base line. The heights of the rectangles are proportional to the frequencies. A typical bar diagram is Figure 2.1.

Example 2.4 In a peach survey on the Murrumbidgee Irrigation Area, Balaam and Corbin (1962) presented the following distribution of yields for 310 plantings of the Golden Queen Cling variety for the 1957–58 season in the Yanco No. 1 Irrigation Area (Table 2.2). For this distribution, find the arithmetic mean and estimate the median. Draw a histogram and a frequency polygon. Calculate the variance of the distribution.

TABLE 2.2   *Frequency distribution of yields (tons/acre)*

Class limits	Frequency	Class limits	Frequency
0·55– 1·55	20	10·55–11·55	18
1·55– 2·55	16	11·55–12·55	19
2·55– 3·55	20	12·55–13·55	14
3·55– 4·55	20	13·55–14·55	12
4·55– 5·55	22	14·55–15·55	13
5·55– 6·55	13	15·55–16·55	10
6·55– 7·55	22	16·55–17·55	5
7·55– 8·55	25	17·55–18·55	5
8·55– 9·55	36	18·55–19·55	4
9·55–10·55	16		
		Total Frequency	310

Assuming these data to be a finite population of peach yields, the arithmetic mean is obtained from the formula,

$$\mu = \sum fX / \sum f$$

Now

$$\sum f = f_1 + f_2 + f_3 + \ldots + f_{17} + f_{18} + f_{19}$$

$$= 20 + 16 + 20 + \ldots + 5 + 5 + 4$$

$$= 310,$$

and

$$\sum fX = f_1 X_1 + f_2 X_2 + \ldots + f_{18} X_{18} + f_{19} X_{19},$$

where X_i is the mid-point of the ith class. Thus

$$X_1 = 0.55 + \tfrac{1}{2}(1.55 - 0.55)$$
$$= 0.55 + 0.50$$
$$= 1.05.$$

Similarly $X_2 = 2.05$, $X_3 = 3.05$ and so on. Therefore

$$\sum fX = 20 \times 1.05 + 16 \times 2.05 + 20 \times 3.05 + \ldots$$
$$+ 5 \times 18.05 + 4 \times 19.05$$
$$= 2615.50,$$

so that $\mu = 2615.50 \div 310$
$$= 8.44 \text{ tons/acre.}$$

The cumulative frequency to the upper class limit (8.55) of the eighth class is 158 (i.e. $20 + 16 + 20 + 20 + 22 + 13 + 22 + 25$). Since the median is the value below which 155 observations lie and above which 155 observations lie, the median is positioned near the upper end of the class 7.55 to 8.55. If an ogive were drawn to position more accurately the median, the value obtained would be the same as that obtained by linear interpolation within the interval 7.55 to 8.55. While formulae exist for positioning the median within this class, an estimate of the median for our purposes may be taken as 8.55.

The histogram and frequency polygon are presented in Figures 2.2

Now
$$(N-1)\sigma^2 = \sum fX^2 - (\sum fX)^2/N$$
$$= \text{total S.S.   (sum of squares).}$$
$$\text{Correction term} = (\sum fX)^2/N$$
$$= (2615.50)^2/310$$
$$= 22067.33,$$
$$\sum fX^2 = 20 \times 1.05^2 + 16 \times 2.05^2 + \ldots$$
$$+ 5 \times 19.05^2 + 4 \times 20.05^2$$
$$= 28682.78,$$

Hence
$$\text{Total S.S.} = 28682.78 - 22067.33$$
$$= 6615.55.$$
$$\text{Variance } \sigma^2 = 6615.55/309$$
$$= 21.410.$$
$$\text{Standard deviation} = \sqrt{21.410} = 4.63 \text{ tons/acre.}$$

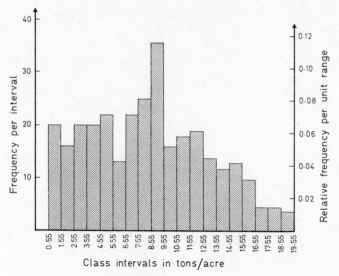

Fig. 2.2   Histogram of peach yields

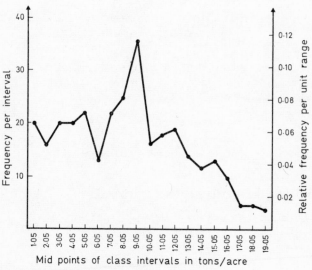

Fig. 2.3   Frequency polygon of peach yields

It is obvious that the preceding arithmetic could not readily be undertaken without a calculating machine.

If the transformation $X = a + cU$, where $a = 9\cdot05$ and $c = 1$ is used the arithmetic is simplified considerably. Using these values,

$$U_1 = X_1 - 9\cdot05 = 1\cdot05 - 9\cdot05 = -8$$
$$U_2 = X_2 - 9\cdot05 = 2\cdot05 - 9\cdot05 = -7$$
$$\vdots$$
$$U_8 = \qquad \dots \qquad = -1$$
$$U_9 = \qquad \dots \qquad = 0$$
$$U_{10} = \qquad \dots \qquad = 1$$
$$\vdots$$
$$U_{19} = \qquad \dots \qquad = 10.$$

The constant a has been chosen so that U_9, the mid-point of the class with the maximum frequency, is the point 0 on the U scale; c has been taken as the class range.

Let $\bar{u}$ be defined as $\sum fU / \sum f$. Then

$$\bar{u} = (f_1 U_1 + f_2 U_2 + \dots + f_{18} U_{18} + f_{19} U_{19})/310$$
$$= (20 \times -8 + 16 \times -7 + \dots + 5 \times 9 + 4 \times 10)/310$$
$$= -190 \div 310$$
$$= -0\cdot61.$$

It should be noted that the largest frequencies are now multiplied by the smallest values of the new variable.

It can be shown that

$$\mu = a + c\bar{u}.$$

Hence

$$\mu = 9\cdot05 + 1 \times (-0\cdot61)$$
$$= 8\cdot44.$$

To obtain the variance of the U variable,

$$\text{Correction term} = (\sum fU)^2/N$$
$$= (-190)^2 \div 310$$
$$= 116\cdot45,$$
$$\sum fU^2 = 20 \times (-8)^2 + 16 \times (-7)^2 + \dots + 5 \times 9^2 + 4 \times 10^2$$
$$= 6732.$$

Hence
$$\text{Total S.S.} = 6732 - 116\cdot45$$
$$= 6615\cdot55$$
$$\sigma_u^2 = 6615\cdot55/309$$
$$= 21\cdot410$$
$$= \sigma_x^2 \quad (\text{since } \sigma_x^2 = c^2\sigma_u^2 \text{ and } c = 1).$$
Standard deviation $= \sqrt{21\cdot410} = 4\cdot63$ tons/acre.

In the above transformation, c was chosen equal to unity since the class interval of X was unity. Where the class interval or class range of X is not unity, it is usual to make c equal the class interval.

The coefficient of variation for this distribution is found as follows :

Coefficient of variation $=$ (standard deviation/mean) $\times 100\%$
$$= (4\cdot63/8\cdot44) \times 100\%$$
$$= 54\cdot9\%.$$

This is a particularly variable population and rarely have data as variable as these to be analyzed.

Example 2.5   The heights of a sample of 100 wheat plants were measured in cm and the following distribution was obtained.

Height (cm)	Frequency	Height (cm)	Frequency
40–49	1	90–99	16
50–59	6	100–109	12
60–69	14	110–119	8
70–79	18	120–129	1
80–89	24		

Find the mean and standard deviation for this sample.

In order that the calculations necessary to find the mean and standard deviation might be systematized, a computation sheet for a frequency distribution is presented in Table 2.3.

The first three columns of the table are prepared from the data in the example. The figures in the U-column are obtained as follows. In the transformation $X = a + cU$, let a and c be chosen so that $U = 0$ when $X = 84\cdot5$ (the mid-point of the modal class) and $U = 1$ when $X = 94\cdot5$. Then a and c can be found, as follows, by substitution.

$$84 \cdot 5 = a + c \cdot 0, \qquad a = 84 \cdot 5,$$
$$94 \cdot 5 = a + c \cdot 1, \qquad c = 10 \cdot 0.$$

Having obtained a and c, the value of U corresponding to $X = 44 \cdot 5$ is found from the equation

$$44 \cdot 5 = 84 \cdot 5 + 10 \cdot 0 U, \qquad U = -4.$$

The other values of U are found in a similar manner.

TABLE 2.3   *Computation sheet*

Height	Mid-point (X) of true interval	f	U	fU	fU^2
40–49	44·5	1	−4	−4	16
50–59	54·5	6	−3	−18	54
60–69	64·5	14	−2	−28	56
70–79	74·5	18	−1	−18	18
80–89	84·5	24	0	0	0
90–99	94·5	16	1	16	16
100–109	104·5	12	2	24	48
110–119	114·5	8	3	24	72
120–129	124·5	1	4	4	16
			Totals	0	296

The entries in the fU and fU^2 columns are now completed.

$$\sum fU = 0$$
$$\bar{u} = 0$$

and since

$$\bar{x} = a + c\bar{u}$$
$$\bar{x} = 84 \cdot 5 + 10 \times 0 = 84 \cdot 5 \text{ cm}$$
$$\left(\sum fU \right)^2 = 0.$$

Correction term, $\left(\sum fU \right)^2 / N = 0,$

$$\sum fU^2 = 296,$$
$$\sum f(U - \bar{u})^2 = 296 - 0 = 296.$$
$$s_u^2 = \left[\sum f(U - \bar{u})^2 \right] / (n - 1)$$
$$= 296 \div 99$$
$$= 2 \cdot 99,$$

and since

$$s_x^2 = c^2 s_u^2,$$
$$s_x^2 = 10^2 \times 2{\cdot}99 = 299,$$
$$s_x = 17{\cdot}3 \text{ cm.}$$

Exercises

2.7   If $\bar{x}. = \sum\limits_{i=1}^{n} X_i/n$ show that

(i) $\sum\limits_{i=1}^{n} (X_i - \bar{x}.) = 0$;

(ii) $\sum\limits_{i=1}^{n} (X_i - a)^2 = \sum\limits_{i=1}^{n} (X_i - \bar{x}.)^2 + n(\bar{x}. - a)^2.$

2.8   If

$$X_i = a + cU_i, \qquad (i = 1, \ldots, n);$$
$$\bar{x}. = \sum X_i/n, \qquad \bar{u}. = \sum U_i/n,$$
$$s_x^2 = \sum (X_i - \bar{x}.)^2/(n-1),$$
$$s_u^2 = \sum (U_i - \bar{u}.)^2/(n-1),$$

show that $\bar{x}. = a + c\bar{u}.$ and $s_x^2 = c^2 s_u^2$.

2.9   (a) Write down in full the terms in the summations

$$\sum_{i=1}^{4} f_i(X_i - \mu)^2, \quad \left(\sum_{i=1}^{4} f_i X_i\right)^2, \quad \sum_{i=1}^{4} f_i X_i^2 .$$

(b) If $\sum\limits_{i=1}^{k} f_i = N$, $\bar{x}. = \sum\limits_{i=1}^{k} f_i X_i/N$ and if $X_i = aU_i + bV_i$ where

a and b are constants, show that $\bar{x}. = a\bar{u}. + b\bar{v}.$ where $\bar{u}.$ and $\bar{v}.$ are defined similarly to $\bar{x}.$.

2.10 For the data given in Table 1.2, let Y_j be the yield of grain of the jth plot and let the plots be numbered as in Exercise 2.5.

(a) Using for $\bar{y}.$ the value $32{\cdot}4$ bushels/acre, prepare a table of

$$(Y_j - \bar{y}.), j = 1, 2, \ldots, 20.$$

Hence find $\sum\limits_{j=1}^{20} (Y_j - \bar{y}.)^2$.

(b) Use the results from Exercise 2.5 to compute

$$\sum_{j=1}^{20} Y_j^2 - \left(\sum_{j=1}^{20} Y_j \right)^2 \Big/ 20.$$

(c) How do the answers in (a) and (b) compare?

2.11   For the frequency distribution of differences in Table 1.4, assuming the data to be a sample of size 300 from an infinite population and that the first and last classes are $-3 \cdot 245$ to $-2 \cdot 995$ and $3 \cdot 005$ to $3 \cdot 255$ respectively, calculate the arithmetic mean. (Obtain the mid-points X_i of the various frequency classes and use the transformation

$$X_i = a + cU_i$$

where $a = -0 \cdot 12$ and $c = 0 \cdot 25$.) Draw the histogram and frequency polygon for this sample. Find the sample variance.

2.12   The following were the yields in bushels per acre of two varieties of oats in five successive years.

| | Annual | yield | | | |
Variety	1	2	3	4	5
A	35	38	30	20	30
B	34	41	25	30	45

Show that the difference between the two varietal means is $4 \cdot 4$ bushels/acre. Calculate the mean of the ten yields and show that the sum of the ten deviations from the general mean is zero.

2.13   For a sample of ten hemp plants the weights in grams were
$$3, 13, 11, 16, 5, 9, 9, 8, 6, 18.$$

Find the sample mean and the standard deviation.

2.14   For the following distribution,
   (a) draw a histogram;
   (b) estimate the median;
   (c) calculate the mean and standard deviation.

Frequency distribution of areas (in arbitrary units) of 200 sheep sperms

Area	Frequency	Area	Frequency
3·6–4·5	3	8·6– 9·5	44
4·6–5·5	8	9·6–10·5	18
5·6–6·5	23	10·6–11·5	4
6·6–7·5	35	11·6–12·5	1
7·6–8·5	63	12·6–13·5	1

2.15   The following table gives the distribution of the yield per plot of 1000 small plots in a wheat uniformity trial.

Class limits grams per plot	Observed frequency
0·05–10·05	129
10·05–20·05	329
20·05–30·05	332
30·05–40·05	102
40·05–50·05	77
50·05–60·05	19
60·05–70·05	6
70·05–80·05	6
Total frequency	1000

Draw a histogram for this distribution, and find the mean and standard deviation of the distribution.

REFERENCES

BALAAM, L. N. and CORBIN, J. B. (1962). 'A Report and Statistical Analysis of Factors Affecting the Yields of Canning Peaches on the Yanco No. 1 Irrigation Area, New South Wales', University of Sydney, School of Agriculture, *Report No. 5*.

FISHER, R. A. (1948). *Statistical Methods for Research Workers* (10th edition). Oliver & Boyd, Edinburgh; Hafner, New York.

COLLATERAL READING

DUNN, Olive Jean (1964). *Basic Statistics: A Primer for the Biomedical Sciences.* John Wiley, New York. Chapter 3.

FINNEY, D. J. (1964). *An Introduction to Statistical Science in Agriculture* (2nd edition). Oliver & Boyd, Edinburgh; John Wiley, New York. Chapter 4.

GOODMAN, R. (1957). *Teach Yourself Statistics.* English Universities Press, London. Chapter 2.

MACK, S. F. (1959). *Elementary Statistics.* Holt, Rinehart and Winston, New York. Chapter 2.

WEATHERBURN, C. E. (1947). *A First Course in Mathematical Statistics.* Cambridge University Press, Cambridge. Chapter 1.

WEILER, H. and WEILER, Gladys E. *Very Elementary Statistics.* William Brooks & Co., Sydney. Chapter 2.

CHAPTER 3

Probability

3.1 Relative Frequency Distributions

When a large finite population is arranged in a frequency distribution, some at least of the classes occur with reasonably large frequencies. In this situation it is often desirable to arrange the data in a *relative frequency distribution*, instead of a frequency distribution.

In the previous chapter the relative frequency (P_i) of the ith class has been defined as f_i/N, where f_i is the frequency of the ith class and N is the total frequency, $\sum f_i$.

Thus, if $P_i = f_i/N$, then $0 \leqslant P_i \leqslant 1$ and

$$\sum P_i = \sum f_i/N = 1,$$

since $\sum f_i = N$. Also, the population mean μ is given by

$$\mu = \sum P_i X_i$$

and the population variance σ^2 is given by

$$\sigma^2 \doteq \sum P_i (X_i - \mu)^2$$
$$= \sum P_i X_i^2 - (\sum P_i X_i)^2$$
$$= \sum P_i X_i^2 - \mu^2,$$

this approximation being very good if N is large. If the variance of a finite population had been defined with a divisor N instead of $N - 1$, the above approximation would have been exact.

Suppose that the population consisted of the number of grains per head of wheat plants and that the modal class X_i occurred with relative frequency P_i. If a head were chosen at random from this population, it would be only seldom that someone could be found who would state that X_i was not the most likely number of grains on this randomly chosen head. Generally, it would be claimed that

41

X_i was the most probable number of grains and that the chance of getting X_i grains was the relative frequency P_i.

Immediately the term relative frequency is used, the threshold of probability theory has been reached, and while two definitions of probability (a mathematical and a statistical definition) will be given, it will be seen that one of these bears a close relationship to the definition of relative frequency. It is in fact the limit of the relative frequency as the total frequency becomes very large.

3.2 Permutations and Combinations

Before the subject of probability is dealt with, a brief review of permutations and combinations will be given. A more detailed account of this branch of algebra will be found in the relevant chapters of any high school algebra textbook.

Considering the four letters a, b, c, d the *permutations* or *arrangements* of these letters two at a time are 12 in number:

$$ab \quad ac \quad ad \quad ba \quad bc \quad bd$$
$$ca \quad cb \quad cd \quad da \quad db \quad dc$$

Permutations are different if they contain the same letters but in a different order.

For n objects taken r at a time, the total number of permutations is given by

$$(n)_r = n(n-1)(n-2)\ldots(n-r+1).$$

If all the n objects are taken, the number of permutations is

$$(n)_n = n(n-1)(n-2)\ldots 3 \times 2 \times 1.$$

The continued product $n(n-1)(n-2)\ldots 3 \times 2 \times 1$ is known as *factorial n* and is written $n!$ (By definition, $0! = 1$.)

If the objects to be arranged are not all different and there are p of the first kind, q of the second, r of the third and s of the fourth, the number of permutations of these objects is

$$\frac{n!}{p!q!r!s!} \quad \text{where } n = p+q+r+s.$$

In *combinations* or *groups*, no account is taken of the order of the objects. Thus the number of combinations of the four letters a, b, c, d

taken two at a time is 6:

$$ab \quad ac \quad ad \quad bc \quad bd \quad cd$$

The two permutations *ab* and *ba* result in only one combination, the group with the letters *a* and *b*.

For *n* objects taken *r* at a time, the number of combinations is

$$\binom{n}{r} = \frac{n(n-1)(n-2)\ldots(n-r+1)}{r!}$$

$$= \frac{n!}{r!\,(n-r)!}\,.$$

3.3 Definitions of Probability

Mathematicians have had lengthy arguments about the two definitions of probability. These need not be the concern of the reader of this text. Both definitions are presented—the mathematical or classical definition first, and the empirical or statistical definition second.

The rolling of an ordinary, cubical die may result in any one of the six different faces facing upward. The group of six possible results is said to be *exhaustive*. The possibility of the die standing on its edge is not considered! The six possible results are said to be *mutually exclusive* since the occurrence of one result precludes the happening of the other. The rolling of the die is called a *trial*.

Suppose that the die is a perfect cube made of homogeneous material and that there is no reason to expect that any particular face will come uppermost more frequently than any other. The six possible results are said to be *equally likely*, or *equally probable*.

Definition. If a trial may result in any one of *n* exhaustive, mutually exclusive and equally likely outcomes, and if *m* of these outcomes entail the occurrence of an event *E*, then the probability that *E* will happen as the result of the trial is given by

$$P(E) = p = m/n\,.$$

Since $0 \leqslant m \leqslant n$, *p* is a positive number less than, or equal to unity.

Since the number of outcomes involving the failure of E is $n - m$, the probability of the failure of E, denoted by $\bar{E}$, is given by

$$P(\bar{E}) = q = (n - m)/n = 1 - m/n = 1 - p.$$

Further,

$$p + q = 1.$$

This method of measuring probability must be confined to those problems in which the outcomes of a trial can be reduced to a certain number of *equally likely* outcomes. Frequently, considerations of symmetry enable a decision to be made in this regard. However, it should be noted that objection is often raised to the above definition of probability because of the use of the idea of *equally probable*.

Can an experimenter be sure that the die is completely symmetrical and that it is unbiased with respect to say the rolling of fives? Surely the only way to test whether this is so is to make a large sequence of trials and observe whether the relative frequency of a five ultimately tends to a ratio of $\frac{1}{6}$.

Thus another definition of the probability of an event is given in terms of the relative frequency of the occurrence of an event in an extended series of trials. The fundamental assumption of such a definition is that this relative frequency, in a uniform series of trials, tends to a definite limit as the number of trials in the series is indefinitely increased. This limit is taken as the measure of the probability of the occurrence of the event in a single trial.

Suppose that out of a series of n trials, the relative frequency of an event E is m/n. Now, it can be shown by actual experimentation that as n increases the relative frequency becomes more consistent. While n is small there will be wide variations in m/n; when n is sufficiently large, it is almost certain that all the values of m/n will be closely grouped around a central value. This central value, which is not a limit in the strict mathematical sense, is the *probability* of E.

As a further clarification of this, the results of an extensive series of trials (rolling a die) might be plotted on a graph as in Figure 3.1. Suppose the event E, whose probability is required, is the showing of a 1 or 2 on the die. In Figure 3.1 the relative frequencies which might be obtained are plotted. It will be seen that as the number of trials n increases, this relative frequency varies only slightly about the ratio $\frac{1}{3}$. The relative frequency is obtained from the formula

$$\text{Relative frequency} = \frac{\text{Number of times die showed 1 or 2}}{\text{Total number of throws of die}}.$$

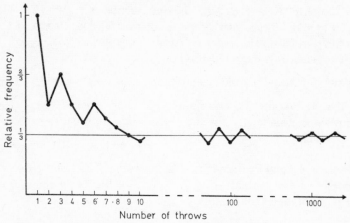

Fig. 3.1   Hypothetical result of an extensive series of trials

Only at certain stages in the experiment will the relative frequency be exactly $\frac{1}{3}$, and if this does happen then immediately there is another throw of the die the relative frequency will not equal $\frac{1}{3}$. For example, suppose that after 99 throws, 33 throws have resulted in a 1 or a 2 being uppermost. The relative frequency is $\frac{33}{99}$ or $\frac{1}{3}$. After the 100th throw, the relative frequency will be either $\frac{33}{100}$ or $\frac{34}{100}$, neither of which is exactly $\frac{1}{3}$. However, the variation about $\frac{1}{3}$ would tend to get smaller and smaller as the number of throws is indefinitely increased.

The difficulty about this definition is this—what grounds are there for assuming that in two different series of experiments, made under exactly the same conditions, the relative frequency would lead to exactly the same limit? No proof can be given that the relative frequency does tend to a limit!

However, depending on the circumstances, there are occasions when one or other of the two definitions is used.

3.4   Theorems of Total and Compound Probability

Rules for computing probabilities of composite or related events in terms of probabilities of simple events (E_i) will now be developed. In the theorems which follow the events 'E_1 or E_2' and 'E_1 and E_2' will be written '$E_1 \cup E_2$', and '$E_1 \cap E_2$' respectively. These are called the 'union of E_1 and E_2' and the 'intersection of E_1 and E_2'. The 'set E_1' may be regarded as the mathematical counterpart of the biologist's real-world 'event E_1'. Just as the + sign is extended to $\sum$, the $\cup$ and $\cap$ are extended so that 'E_1 or E_2 or E_3 or ... or E_k' is written $\bigcup\limits_{i=1}^{k} E_i$ and 'E_1 and E_2 and E_3 and ... and E_k' is written $\bigcap\limits_{i=1}^{k} E_k$. While these terms and this notation is taken from set theory, no knowledge of set theory is required to understand what is contained in this section.

Theorem of Total Probability. The probability that some one or other of several mutually exclusive events will occur is the sum of the probabilities of the separate events.

Suppose that a trial may result in any one of n equally probable outcomes, of which m_1 entail the occurrence of an event E_1, m_2 the occurrence of $E_2, \ldots, m_k$ the occurrence of E_k. If the events $E_1, E_2, \ldots, E_k$ are mutually exclusive, the corresponding outcomes are all different. The number of outcomes which entail the occurrence of either E_1 or $E_2 \ldots$ or E_k is $m_1 + m_2 + \ldots + m_k$. Hence the probability that one of these events will occur is

$$P\left(\bigcup_{i=1}^{k} E_i\right) = (m_1 + m_2 + \ldots + m_k)/n$$

$$= (m_1/n) + (m_2/n) + \ldots + (m_k/n)$$

$$= \sum_{i=1}^{k} p_i$$

where p_i is the probability of the event E_i.

Definition. An event E_2 is *dependent* on an event E_1 if the occurrence of E_1 or its non-occurrence alters the probability of E_2. If, however, neither the occurrence nor the non-occurrence of E_1 affects the probability of E_2, E_2 is *independent* of E_1.

Suppose that in a plot of wheat plants there are d diseased and h healthy plants. Consider the random selection of two plants in succession. The finding of the first plant to be diseased will be event E_1; the finding of the second plant to be diseased will be event E_2.

The probability of E_1 is $d/(d+h)$. However the probability of E_2 depends on whether E_1 has occurred or not. If E_1 has occurred the probability of E_2 is $(d-1)/(d+h-1)$ whereas if E_1 has not occurred, the probability of E_2 is $d/(d+h-1)$.

The first of these, $(d-1)/(d+h-1)$, is the *conditional probability* of E_2 on the assumption that E_1 has occurred, and is written

$$P(E_2 | E_1).$$

The second, $d/(d+h-1)$, is the conditional probability of E_2 on the assumption that E_1 has not occurred and is written

$$P(E_2 | \bar{E}_1).$$

More generally, suppose that a trial may result in any one of n equi-probable outcomes, some of which entail the occurrence of E_1 alone, some that of E_2 alone, some that of both E_1 and E_2 and some neither E_1 nor E_2. Let m_1 be the number of outcomes which entail the occurrence of E_1. The number of outcomes which entail the occurrence of both E_1 and E_2 is included in this number. Let m ($m \leqslant m_1$) be their number. Then the probability that both E_1 and E_2 will occur is

$$P(E_1 \cap E_2) = P(E_2 \cap E_1)$$

$$= m/n$$

$$= \frac{m}{m_1} \cdot \frac{m_1}{n}$$

$$= P(E_2 | E_1)P(E_1).$$

Since m is the number of outcomes which entail E_2 as well as E_1, m/m_1 is the conditional probability of E_2 on the assumption that E_1 has happened. The quotient m_1/n is the probability of the event E_1.

Theorem of Compound Probability. The probability of the combined occurrences of two events E_1 and E_2, is the product of the probability of E_1 and the conditional probability of E_2 on the assumption that E_1 has happened, i.e.

$$P(E_1 \cap E_2) = P(E_1)P(E_2 | E_1).$$

When the events E_1 and E_2 are independent, the result may be stated simply that the probability that two independent events will both happen is the product of the probabilities of the separate events, i.e.

$$P(E_1 \cap E_2) = P(E_1)\,P(E_2).$$

This theorem may be extended to include any number of events, e.g.

$$P(E_1 \cap E_2 \cap E_3) = P(E_1 \mid E_2 \cap E_3)P(E_2 \mid E_3)P(E_3)$$

if dependent and

$$P(E_1 \cap E_2 \cap E_3) = P(E_1)P(E_2)P(E_3)$$

if independent. For the general independent case

$$P\left(\bigcap_{i=1}^{k} E_i\right) = \prod_{i=1}^{k} P(E_i).$$

Note: The $\prod$ (pi) notation is similar to the $\sum$ notation. While $\sum$ stands for 'summation', $\prod$ stands for 'product'.

3.5   Venn Diagrams

The ideas underlying composite events and their probabilities can be presented graphically by means of Venn diagrams shown in Figure 3.2. These diagrams are interpreted as follows:

The total number of equally probable outcomes is represented by the square area of the diagram. The number of equally likely outcomes which entail the occurrence of E_1 is represented by the area within the loop marked E_1 or, in the case of Figure 3.2(b), by the area within the rectangle $APQD$. Similarly, the number of equally likely outcomes which entail the occurrence of E_2 is represented by the area within the loop E_2 or the rectangle $LNCD$.

In Figure 3.2(a), E_1 and E_2 are mutually exclusive events and

$$\begin{aligned} P(E_1 \cup E_2) &= P(E_1 \text{ or } E_2) \\ &= P(E_1) + P(E_2). \end{aligned}$$

In Figure 3.2(b), E_1 and E_2 are independent events; the double hatched area $LMQD$ represents those outcomes favourable to both E_1 and E_2 and

$$\begin{aligned} P(E_1 \mid E_2) &= \text{area } LMQD/\text{area } LNCD \\ &= \text{area } APQD/\text{area } ABCD \\ &= P(E_1). \end{aligned}$$

Also,

$$P(E_1 \cap E_2) = P(E_1 \text{ and } E_2)$$

$$= \text{area } LMQD/\text{area } ABCD$$

$$= \frac{\text{area } LMQD}{\text{area } LNCD} \cdot \frac{\text{area } LNCD}{\text{area } ABCD}$$

$$= \frac{\text{area } APQD}{\text{area } ABCD} \cdot \frac{\text{area } LNCD}{\text{area } ABCD}$$

$$= P(E_1)P(E_2).$$

As in Figure 3.2(b), the double hatched areas in Figure 3.2(c) and (d) represent those outcomes which entail the occurrence of both

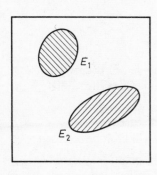

(a) Mutually exclusive events

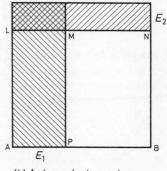

(b) Independent events

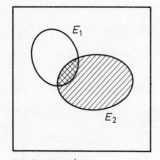

(c) Event $E_1|E_2$

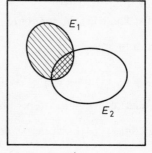

(d) Event $E_2|E_1$

Fig. 3.2　Venn diagrams

E

E_1 and E_2. Figures 3.2(c) and (d) are the Venn diagrams when the events are not independent. In Figure 3.2(c)

$$P(E_1 | E_2) = P(E_1 \cap E_2)/P(E_2),$$

while in Figure 3.2(d)

$$P(E_2 | E_1) = P(E_2 \cap E_1)/P(E_1).$$

Example 3.1 What is the probability of throwing an even number with a die?

The six faces numbered $1, 2, 3, 4, 5, 6$ are six exhaustive mutually exclusive and equally likely outcomes. Let the event E be the throwing of an even number. The number of outcomes which entail the occurrence of E is three, i.e. the throwing of a 2, or a 4, or a 6. Hence,

$$P(E) = \frac{m}{n} = \frac{3}{6} = \frac{1}{2}.$$

Example 3.2 In a small plot of wheat there are twenty plants. Of these, five are diseased and the remainder healthy. What is the probability

(i) that a plant selected at random is diseased;
(ii) that two plants selected at random will both be diseased.

(i) The trial, the random selection of a plant, may result in any one of twenty exhaustive, mutually exclusive and equally likely outcomes and of these five outcomes entail the occurrence of the event E, the selection of a diseased plant. Therefore, the required probability is 5/20.

(ii) The trial, the random selection of a group of two plants, may result in any one of $\binom{20}{2}$ exhaustive, mutually exclusive and equally likely outcomes. $\binom{20}{2}$ is the number of groups of two which may be selected from the twenty plants in the plot. Of the total number of outcomes, $\binom{5}{2}$ outcomes entail the occurrence of the event E, the selection of two diseased plants. $\binom{5}{2}$ is the number of groups of two diseased plants which may be selected from the five diseased plants in the plot. Thus

$$P(E) = \frac{m}{n} = \binom{5}{2} \Big/ \binom{20}{2}$$
$$= \frac{5.4}{2.1} \cdot \frac{2.1}{20.19} = \frac{1}{19}.$$

This probability could also be calculated using the theorem of compound probability.

Let E_1 be the event—'first plant chosen, diseased' and let E_2 be the event—'second plant chosen, diseased'. Then

$$P(E_1) = \frac{\text{No. of diseased plants in plot}}{\text{Total No. of plants in plot}} = \frac{5}{20},$$

$$P(E_2 | E_1) = \tfrac{4}{19},$$

$$P(E_1 \cap E_2) = P(E_1)P(E_2 | E_1)$$
$$= \tfrac{5}{20} \cdot \tfrac{4}{19} = \tfrac{1}{19}.$$

Example 3.3 A simple genetic example illustrating the use of the theorem of total probability is the following. Suppose that the three types WW, Ww, ww occur with probabilities p, $2q$ and r where

$$p + 2q + r = 1$$

and that there is complete dominance. Then the probability of an individual drawn at random being dominant (i.e. type WW or Ww) is

$$P(WW \cup Ww) = P(WW) + P(Ww)$$
$$= p + 2q.$$

Example 3.4 Three cages of insects contain respectively 30 insects of Strain A and 10 of Strain B, 20 of Strain A and 20 of B, 10 of A and 30 of B. One insect is chosen at random from each cage. What is the probability that the three selected insects comprise 1 of A and 2 of B?

The event E may occur in any one of the following mutually exclusive ways:

(i) A from first cage, B from second cage, B from third cage.

(ii) B from first cage, A from second cage, B from third cage.
(iii) B from first cage, B from second cage, A from third cage.

The probabilities required in (i), (ii), and (iii) may be obtained by the theorem of compound probability and since the selection of the insects from the three cages are independent,

$$P(i) = P(A \text{ from first cage})P(B \text{ from second cage})P(B \text{ from third cage})$$

$$= \tfrac{30}{40} \cdot \tfrac{20}{40} \cdot \tfrac{30}{40} = \tfrac{9}{32}.$$

Similarly,

$$P(ii) = \tfrac{10}{40} \cdot \tfrac{20}{40} \cdot \tfrac{30}{40} = \tfrac{3}{32},$$

$$P(iii) = \tfrac{10}{40} \cdot \tfrac{20}{40} \cdot \tfrac{10}{40} = \tfrac{1}{32}.$$

Since the events (i), (ii) and (iii) are mutually exclusive then, by the theorem of total probability,

$$P(E) = P(i) + P(ii) + P(iii)$$

$$= \tfrac{9}{32} + \tfrac{3}{32} + \tfrac{1}{32}$$

$$= \tfrac{13}{32}.$$

Example 3.5   Assume that the probability is $\tfrac{1}{2}$ that a calf at birth is a heifer, that a cow in four consecutive parturitions has four single-born calves and that the sex of a calf at a particular parturition is independent of the sex at a previous parturition. What is the probability that the cow

(a) has four heifer-calves;
(b) has three heifer-calves followed by a bull-calf;
(c) has three heifer-calves and a bull-calf.

(a) Let

E_1 be the event that the first calf is a heifer;
E_2 be the event that the second calf is a heifer;
E_3 be the event that the third calf is a heifer;
E_4 be the event that the fourth calf is a heifer.

Then

$$P(E_i) = \tfrac{1}{2}, \qquad (i = 1, 2, 3, 4).$$

Now the event 'four heifer-calves' is the event E_1 and E_2 and E_3 and E_4.

By the theorem of compound probability

$$P\left(\bigcap_{i=1}^{4} E_i\right) = \prod_{i=1}^{4} P(E_i)$$

$$= \tfrac{1}{2} \cdot \tfrac{1}{2} \cdot \tfrac{1}{2} \cdot \tfrac{1}{2} = \tfrac{1}{16}.$$

(b) Let $\bar{E}_4$ be the event that the fourth calf is a bull. Then $P(\bar{E}_4) = \tfrac{1}{2}$ and, similarly to (a),

$$P(E_1 \cap E_2 \cap E_3 \cap \bar{E}_4) = (\tfrac{1}{2})^4 = \tfrac{1}{16}.$$

(c) The event 'three heifer-calves and a bull-calf' is different from that in (b). Here the bull-calf might be the first followed by three heifers or it might be the second, or the third or the fourth. Let

A_1 be the composite event that the first calf is a bull; the second, a heifer; the third, a heifer and the fourth, a heifer;
A_2 be the composite event that the first calf is a heifer; the second, a bull; the third, a heifer and the fourth, a heifer;
A_3 and A_4 be similarly defined.

Then by the theorem of compound probability,

$$P(A_i) = \tfrac{1}{16}, \qquad (i = 1, 2, 3, 4).$$

The event 'three heifer-calves and a bull-calf' will occur if A_1 or A_2 or A_3 or A_4 occurs. Thus by the theorem of total probability

$$P\left(\bigcup_{i=1}^{4} A_i\right) = \sum_{i=1}^{4} P(A_i)$$

$$= \tfrac{1}{16} + \tfrac{1}{16} + \tfrac{1}{16} + \tfrac{1}{16} = \tfrac{1}{4}.$$

Example 3.6 Suppose that when a true-breeding purple-stem potato-leaf strain of tomatoes is crossed with a true-breeding green-stem cut-leaf strain, the following four classes occur:– purple, cut (AB); purple, potato $(A\bar{B})$; green, cut $(\bar{A}B)$; green, potato $(\bar{A}\bar{B})$. If the events A and B (and hence $\bar{A}$ and $\bar{B}$) are independent and the probabilities of $A, \bar{A}, B, \bar{B}$, are respectively $\tfrac{3}{4}, \tfrac{1}{4}, \tfrac{3}{4}, \tfrac{1}{4}$, then by the theorem of compound probability,

$$P(A \cap B) = P(A)P(B)$$

$$= \tfrac{3}{4} \times \tfrac{3}{4} = \tfrac{9}{16},$$

$$P(A \cap \bar{B}) = P(A)P(\bar{B})$$
$$= \tfrac{3}{4} \times \tfrac{1}{4} = \tfrac{3}{16},$$

$$P(\bar{A} \cap B) = P(\bar{A})P(B)$$
$$= \tfrac{1}{4} \times \tfrac{3}{4} = \tfrac{3}{16},$$

$$P(\bar{A} \cap \bar{B}) = P(\bar{A})P(\bar{B})$$
$$= \tfrac{1}{4} \times \tfrac{1}{4} = \tfrac{1}{16},$$

The events for which probabilities have been calculated could be represented by a Venn diagram similar to Figure 3.2(b) (p. 49). If the sides DC and DA were each of length 4 cm and DQ and DL were each of length 3 cm, then $DQML$ would be of area 9 cm^2 and represent the probability of $A \cap B$ which is $\tfrac{9}{16}$. Similarly, $QCLM$, whose area would be 3 cm^2, would represent the probability of $A \cap \bar{B}$; $LMPA$, the probability of $\bar{A} \cap B$; and $MNBP$, the probability of $\bar{A} \cap \bar{B}$.

Exercises

3.1 What is the probability that in rolling a die a multiple of three will be uppermost?

3.2 What is the probability of obtaining a total of nine points in a single throw of two dice?

3.3 The blood classifications in a class of 100 biometry students are 28 A's, 12 B's, 17 AB's, 43 O's. Given that the probability that two students selected at random will both have blood type A is $\binom{28}{2} \Big/ \binom{100}{2}$, what is the probability that two students selected at random will both have the same blood classification?

3.4 To each of ten adjacent plots in a row, either fertilizer A or fertilizer B is to be allocated. The allocation is random, the process being—'for each plot a coin is tossed and if the result is a head, fertilizer A is applied to the plot, if tails, fertilizer B'. What is the probability that

(a) Fertilizer B is not applied to any plot?

(b) Fertilizer A is applied to exactly two plots?

(c) Fertilizer A is applied to the first five plots and fertilizer B to the next five plots?

(d) Fertilizers A and B or B and A are applied alternately?

3.5 6 treatments (numbered 1 to 6) are to be allocated to 24 plots (numbered 1 to 24) by rolling a die 24 times; the number uppermost on the first roll being the treatment allotted to the first plot and so on.

(a) What is the probability that the allocation of treatments is as shown?

Plot No.	1	2	3	4	5	6	7	8	9	10	11	12
Treat No.	1	1	1	1	2	2	2	2	3	3	3	3

Plot No.	13	14	15	16	17	18	19	20	21	22	23	24
Treat No.	4	4	4	4	5	5	5	5	6	6	6	6

(b) What is the probability that the allocation of treatments is as shown?

Plot No.	1	2	3	4	5	6	7	8	9	10	11	12
Treat No.	1	6	5	2	4	4	3	1	5	6	2	3

Plot No.	13	14	15	16	17	18	19	20	21	22	23	24
Treat No.	4	5	1	2	3	6	5	4	4	3	6	1

3.6 Discrete Probability Distributions

Suppose that, corresponding to the k exhaustive, mutually exclusive outcomes in a trial, a variable X takes the k values X_i with probability p_i, $(i = 1, 2, \ldots, k)$, i.e.

$$P(X = X_i) = p_i.$$

For example, in the throw of a die, corresponding to the six exhaustive, mutually exclusive outcomes, the variable X takes the six values $1, 2, 3, 4, 5, 6$ each with probability $\frac{1}{6}$.

A *probability distribution* is defined to be the assemblage of all possible values X_i which a variable X may take, together with their associated probabilities, p_i. Thus for a probability distribution, $\sum p_i = 1$. A variable distributed in a probability distribution will be called a *variate*. Most of the concepts associated with relative frequency distributions are applicable to probability distributions. The definitions of the mean and variance of a probability distribution are similar to those already defined for frequency distributions.

The mean of a probability distribution is called the *expected value* or *expectation*. It is denoted by $E(X)$ and is defined as

$$\mu = E(X) = \sum_{i=1}^{k} p_i X_i.$$

Since probability p has been defined as the limit as n approaches infinity of the ratio m/n, then p_i is the limit of the relative frequency f_i/N (or P_i). Thus the probability distribution is the limiting form, as the total frequency approaches infinity, of the relative frequency distribution.

Further, for a sample of size N where

$$\bar{x} = \frac{1}{N} \sum_{i=1}^{k} f_i X_i$$

$$= \sum_{i=1}^{k} (f_i/N) X_i = \sum_{i=1}^{k} P_i X_i$$

it is seen that the expected value of X, $E(X)$, is the limit of $\bar{x}$.

The probability distribution of X^2 is the assemblage of values X_i^2 with the probabilities p_i, where p_i is the probability of the value X_i.

It will be appreciated that if the X_i may be positive or negative, the set of X_i^2 may have values which occur twice. For example if X takes the values $-2, -1, 0, 1, 2$ with probabilities $0.1, 0.2, 0.4, 0.2, 0.1$, the corresponding set of X_i^2 is $4, 1, 0, 1, 4$. The probability distribution of X^2 is

$$X_i^2: \quad 4, \quad 1, \quad 0, \quad 1, \quad 4.$$
$$p_i: 0.1, 0.2, 0.4, 0.2, 0.1.$$

For this distribution, the expected value of the variate X^2 is

$$E(X^2) = \sum p_i X_i^2.$$

Returning to the distribution of X, the variance of this distribution is defined by

$$\sigma^2 = E[X - E(X)]^2.$$

This variance is frequently computed from the formula

$$\sigma^2 = E(X^2) - [E(X)]^2.$$

Not many lines of algebra are needed to show that the definitive and computation formulae are equal. Further, they can be shown to be the limiting form (as the sample size becomes infinitely large) of the sample variance

$$s^2 = \sum_{i=1}^{k} f_i (X_i - \bar{x})^2 / (n-1).$$

Further, in a manner similar to that in which the sum of deviations from the arithmetic mean of a frequency distribution can be shown to be zero, the expected value of the deviation of a variate from its mean can be shown to be zero, i.e.

$$E[X - E(X)] = 0.$$

3.7 Some Important Theorems

The following five theorems are used frequently in biometry. The first theorem is analogous to the results presented in section 2.8 concerning linear transformations of variables. Suppose a biometrical analysis is made of a variable (yield) measured in pounds per acre (or per plot) and it becomes necessary to present the calculated statistics (mean and variance) in units of bushels per acre. Theorem 3.1 would be used in instances such as this.

Theorems 3.2 to 3.5 are important theoretically and will be used on many occasions throughout this text. However, their proofs are not presented because of the tedious algebra which is involved.

Theorem 3.1 If the variate X is distributed with mean $\mu = E(X)$ and variance

$$\sigma^2 = E(X^2) - [E(X)]^2,$$

then

$$T = cX + d$$

where c and d are constants is distributed with mean $c\mu + d$ and variance $c^2 \sigma^2$.

$$
\begin{aligned}
\text{Proof. (i)} \quad E(T) &= E(cX + d) \\
&= \sum p_i (cX_i + d) \\
&= c \sum p_i X_i + d \sum p_i \\
&= c\mu + d \quad \text{since } \sum p_i = 1.
\end{aligned}
$$

Now
$$\text{(ii)} \quad \text{var}(T) = E(T^2) - \{E(T)\}^2.$$

$$
\begin{aligned}
E(T^2) &= E(cX+d)^2 \\
&= E(c^2 X^2 + 2cdX + d^2) \\
&= \sum p_i(c^2 X_i^2 + 2cdX_i + d^2) \\
&= c^2 \sum p_i X_i^2 + 2cd\mu + d^2.
\end{aligned}
$$

Again
$$
\begin{aligned}
[E(T)]^2 &= (c\mu+d)^2 \\
&= c^2\mu^2 + 2cd\mu + d^2.
\end{aligned}
$$

Thus
$$
\begin{aligned}
\text{var}(T) &= E(T^2) - \{E(T)\}^2 \\
&= c^2 \sum p_i X_i^2 - c^2\mu^2 \\
&= c^2\{\sum p_i X_i^2 - (\sum p_i X_i)^2\} \\
&= c^2\sigma^2.
\end{aligned}
$$

Corollary (i) $U = X + d$ is distributed with mean $\mu + d$ and variance σ^2;

(ii) $V = cX$ is distributed with mean $c\mu$ and variance $c^2\sigma^2$.

Theorem 3.2 The expected value of the sum (or difference) of two variates X and Y is equal to the sum (or difference) of their expected values, i.e.

$$E(X+Y) = E(X) + E(Y).$$

This theorem can be extended to any number of variates. Thus

$$E(X+Y+Z) = E(X) + E(Y) + E(Z)$$

or, more generally, if $V_i(i = 1, 2, 3, \ldots, v)$ are v variates

$$E\left(\sum_{i=1}^{v} V_i\right) = \sum_{i=1}^{v} E(V_i).$$

It should be noted that in this theorem no reference is made to dependence or otherwise of the variates.

Theorem 3.3 The expected value of the product of two independent variates X and Y is the product of their expected values, i.e.

$$E(XY) = E(X)E(Y).$$

On extending to v independent variates, V_i,

$$E\left(\prod_{i=1}^{v} V_i\right) = \prod_{i=1}^{v} E(V_i).$$

Definition. The *covariance* of two variates is the expected value of the product of the deviations of the two variates from their respective expected values, i.e.

$$\begin{aligned} \text{cov}(X, Y) &= E[\{X - E(X)\}\{Y - E(Y)\}] \\ &= E(XY) - E(X)E(Y). \end{aligned}$$

This formula is analogous to that for the variance, i.e.

$$\begin{aligned} \text{var}(X) &= E[\{X - E(X)\}\{X - E(X)\}] \\ &= E(X^2) - [E(X)]^2. \end{aligned}$$

Corollary. The covariance of two independent variates is zero since

$$\begin{aligned} \text{cov}(X, Y) &= E(XY) - E(X)E(Y) \\ &= E(X)E(Y) - E(X)E(Y) \\ [E(XY) &= E(X)E(Y) \text{ because independent}] \\ &= 0. \end{aligned}$$

Theorem 3.4 The variance of the sum (or difference) of two variates X and Y is equal to the sum of their variances together with (or diminished by) twice their covariance, i.e.

$$\begin{aligned} \text{var}(X + Y) &= \text{var}(X) + \text{var}(Y) + 2\,\text{cov}(X, Y) \\ \text{var}(X - Y) &= \text{var}(X) + \text{var}(Y) - 2\,\text{cov}(X, Y). \end{aligned}$$

Since the covariance of independent variates is zero, this theorem may be extended to the following theorem.

Theorem 3.5 The variance of the sum (or difference) of two independent variates X and Y is equal to the sum of their variances, i.e.

$$\text{var}(X + Y) = \text{var}(X) + \text{var}(Y)$$

or

$$\text{var}(X - Y) = \text{var}(X) + \text{var}(Y).$$

The section of this theorem relating to the variance of the sum may be extended to any number of independent variates, e.g. for independent variates, Z_i,

$$\text{var} \sum_{i=1}^{n} Z_i = \sum_{i=1}^{n} \text{var}(Z_i).$$

The first reaction by many students to Theorem 3.5 is that it is false. How can the variance of the sum of two variables be the same as the variance of the difference? This reaction comes about because we are accustomed in real life to consider variables which are not independent (e.g. yields of grain in two adjacent fields; weight gains by animals in different periods of time) and because we tend to think of large numbers having large variability, small numbers, small variability.

To show that first impressions can be misleading, consider the following artificial example. Suppose $18, 19, 20, 21, 22$ are the yields from five treatment-A plots while $27, 24, 23, 24, 27$ are the yields from five treatment-B plots. The sums of yields are

$$18 + 27 = 45; \quad 19 + 24 = 43; \quad 20 + 23 = 43; \quad 21 + 24 = 45;$$
$$22 + 27 = 49.$$

The variance of $45, 43, 43, 45, 49$ is 6 if a divisor of 4 is used. The differences are

$$18 - 27 = -9; \quad 19 - 24 = -5; \quad 20 - 23 = -3; \quad 21 - 24 = -3;$$
$$22 - 27 = -5.$$

The variance of $-9, -5, -3, -3, -5$ is also 6 if a divisor of 4 is used. The variances of the sums and differences are the same. This is so because the two sets of yields were chosen to be independent. If the yields for one of the treatments were rearranged, so that the two sets were no longer independent, the variance of the sums would not equal the variance of the differences.

The variance of the mean yield for a treatment has to be calculated frequently in biometry. Suppose the variance of each of the five A yields is σ^2 and that the yields are independent. Then the mean yield is

$$(18 + 19 + 20 + 21 + 22)/5 = 20.$$

Theorem 3.5 asserts that the variance of $(18 + 19 + 20 + 21 + 22)$ is

$$\sigma^2 + \sigma^2 + \sigma^2 + \sigma^2 + \sigma^2 = 5\sigma^2.$$

Since the total yield is now divided by 5 to obtain the mean, Theorem 3.1 states that the variance of 20 (the mean) is $(\frac{1}{5})^2$ times the variance of the total. Thus the variance of the mean is

$$5\sigma^2/25 = \sigma^2/5.$$

(This is an illustration of a theorem which will be considered in detail in Chapter 6, which states that the variance of the mean of a random sample of size n is σ^2/n.)

Exercises

3.6 In a study on the relationship between days to head (X) and days to mature (Y) for wheat the probabilities for the various pairs (X, Y) are given in Table 3.1. From this table compute $E(X), E(Y), E(X^2), E(Y^2), E(XY)$, variance of X, variance of Y and the covariance of X and Y. Discuss the independence, or otherwise, of X and Y.

TABLE 3.1 *Probabilities of* (X_i, Y_j)

Days to mature (Y_j)	Days to Head (X_i)								
	53	54	55	56	57	58	59	60	61
88	0·01	0·01	0·01						
89	0·01	0·02	0·04	0·02	0·01				
90		0·02	0·04	0·04	0·03	0·02			
91		0·01	0·03	0·08	0·06	0·03	0·01		
92			0·01	0·05	0·07	0·05	0·03	0·01	
93				0·01	0·04	0·04	0·04	0·02	
94						0·03	0·04	0·02	0·01
95							0·01	0·01	0·01

3.7 Let U and V be two independently distributed variates with variances σ_u^2 and σ_v^2 respectively.
What is the variance of:

(i) the distribution of $U - V$;
(ii) the distribution of $U + V$;
(iii) the distribution of $2U - 3V$?

3.8 It is known that the yields of grain (Y) from small plots are distributed with mean 1200 lb/acre and standard deviation 120 lb/acre. The yields are converted to bushels/acre by the transformation $B = Y/60$. What are the mean and variance of the variable B?

3.9   An experiment is designed to study the yielding ability of two varieties, A and B. The two varieties are grown in a number of pairs of adjacent plots; variety A is grown in one plot of the pair, variety B in the other. Suppose that the yields (X) of variety A are distributed with mean 25 bushels/acre and variance 36 (bushels/acre)^2 and that the yields (Y) of variety B are distributed with mean 20 bushels/ acre and variance 25 (bushels/acre)^2. Suppose further that it is known that the yield of one plot is related to the yield of the plot adjacent to it, that this relationship is measured by the covariance in yields of adjacent plots, and that the covariance of adjacent plot yields in this experiment is 20 (bushels/acre)^2. The variable of interest to the experimenter is the difference, D, in yields of A and B plots; $(D = X - Y)$. Use Theorem 3.2 to find the mean of D and Theorem 3.4 to calculate the variance of D.

COLLATERAL READING

MACK, S. F. (1960). *Elementary Statistics*. Holt Rinehart and Winston, New York. Chapter 3.

WEATHERBURN, C. E. (1947). *A First Course in Mathematical Statistics*. Cambridge University Press, Cambridge. Chapter 2.

CHAPTER 4

Two Discrete Probability Distributions

4.1 The Binomial Distribution

In this chapter two discrete probability distributions which are frequently used by biologists are considered. In order to develop the theory associated with these distributions recourse will be made to some of the results of the previous chapter. The theorems of total and compound probability and the formulae for the expected value and variance of a probability distribution will be used. In the form in which the theorem of compound probability is used, it is well to remember that the events are assumed to be independent. Independence then is an implicit assumption underlying both the binomial and Poisson distributions which are now about to be studied.

Suppose that the occurrence of an event E (and hence its non-occurrence $\bar{E}$) is studied in a series of n independent trials, and suppose further that the probability of success (the occurrence of E) is the same for each trial. Suppose this probability is π. Then the probability of $\bar{E}$ is $\delta = 1 - \pi$.

The event E may be the occurrence of a diseased plant and $\bar{E}$ the occurrence of a healthy plant. The n independent trials may be the examination of one hundred plants chosen at random from a field. Again, the event E might be that an animal is male, or that a flower is a particular colour, or that a seed has a certain form or shape, or that an insect has survived. In many experiments when observations such as these are made, the properties of the binomial distribution are useful in interpretating the data. The binomial distribution is sometimes called the Bernoulli distribution after J. Bernoulli, who first described this distribution towards the end of the seventeenth century.

The variable to be studied is the number of times (X) that E occurs in the series of n trials. X is a discrete variable and may take the values $0, 1, 2, \ldots, n$. The probability that $X = r$ is found as follows.

Since $X = r$, the number of failures $= n - r$.

By the theorem of compound probability, the probability of r successes and $n - r$ failures in a specified order is

$$(\pi)^r (1 - \pi)^{n-r}.$$

However, the number of different orders in which these r successes and $n - r$ failures may occur in the n trials is the number of ways of selecting r out of the n positions for the successes. This is $\binom{n}{r}$.

Thus, by the theorem of total probability, the probability of exactly r successes in the series of n independent trials where π is the probability of success in a trial is given by

$$P(X = r) = \binom{n}{r}(\pi)^r(1 - \pi)^{n-r}.$$

This is the *probability function* for the binomial distribution.

The probability that X takes one of the values $0, 1, 2, \ldots, n$ is obtained by substituting for r in the above formula. Thus

$$P(X = 0) = \binom{n}{0}\pi^0(1 - \pi)^n$$
$$= (1 - \pi)^n,$$
$$P(X = 1) = \binom{n}{1}\pi^1(1 - \pi)^{n-1},$$
$$P(X = 2) = \binom{n}{2}\pi^2(1 - \pi)^{n-2},$$
$$\vdots \qquad \qquad \vdots$$
$$P(X = n) = \binom{n}{n}\pi^n(1 - \pi)^{n-n}$$
$$= \pi^n.$$

This distribution is known as the binomial distribution because the various probabilities can be found from the terms in the binomial expansion, $(\delta + \pi)^n$. The binomial distribution is a family of distributions; for different values of n and/or π, different members of the family are obtained. The binomial distribution is said to have two parameters, n and π.

Example 4.1 Suppose that a red shorthorn bull is mated with roan cows on ten occasions. What are the probabilities that the number of red calves will be $0, 1, 2, \ldots 9, 10$, given that there is a probability of 0.5 that a calf from a roan cow mated to a red bull will be red.

Let X be the number of red calves. Then

$$P(X = r) = \binom{n}{r} \pi^r (1-\pi)^{n-r}$$

$$= \binom{10}{r} 0.5^r 0.5^{n-r} \qquad \text{since } \pi = 0.5$$

$$= \binom{10}{r} 0.5^{10} \qquad \text{since } n = 10$$

$$= \binom{10}{r} \bigg/ 1024 \qquad \text{since } 0.5^{10} = 1/1024.$$

r	$P(X = r)$	
0	$\binom{10}{0} \big/ 1024 = \dfrac{1}{1024}$	$= 0.001$
1	$\binom{10}{1} \big/ 1024 = \dfrac{10}{1} \times \dfrac{1}{1024}$	$= 0.010$
2	$\binom{10}{2} \big/ 1024 = \dfrac{10 \times 9}{2 \times 1} \times \dfrac{1}{1024}$	$= 0.044$
3	$\binom{10}{3} \big/ 1024 = \dfrac{10 \times 9 \times 8}{3 \times 2 \times 1} \times \dfrac{1}{1024}$	$= 0.117$
4	$\binom{10}{4} \big/ 1024 = \dfrac{10 \times 9 \times 8 \times 7}{4 \times 3 \times 2 \times 1} \times \dfrac{1}{1024}$	$= 0.205$
5	$\binom{10}{5} \big/ 1024 = \dfrac{10 \times 9 \times 8 \times 7 \times 6}{5 \times 4 \times 3 \times 2 \times 1} \times \dfrac{1}{1024} = 0.246$	
6	$\binom{10}{6} \big/ 1024 = \binom{10}{4} \big/ 1024$	$= 0.205$

F

r	$P(X = r)$	
7	$\binom{10}{7}\Big/1024 = \binom{10}{3}\Big/1024$	$= 0{\cdot}117$
8	$\binom{10}{8}\Big/1024 = \binom{10}{2}\Big/1024$	$= 0{\cdot}044$
9	$\binom{10}{9}\Big/1024 = \binom{10}{1}\Big/1024$	$= 0{\cdot}010$
10	$\binom{10}{10}\Big/1024 = \binom{10}{0}\Big/1024$	$= 0{\cdot}001$

In the table above, the probabilities of $X = 6, 7, 8, 9, 10$ have been calculated using the formula

$$\binom{n}{r} = \binom{n}{n-r}.$$

It should be noted that the distribution is symmetrical. This is true of all binomial distributions where $\pi = 0{\cdot}5$.

It might also be noted that the sum of the probabilities is unity, as it should be, since the probabilities are the terms of the expansion $(\frac{1}{2} + \frac{1}{2})^{10}$.

This illustrates what was stated in section 3.6, i.e. $\sum p_i = 1$. That is, in a probability distribution the probabilities associated with the different values (X_i), which the variable X may assume, sum to unity.

Example 4.2 For Example 4.1, what is the probability that the number of red calves will be three or more?

For all probability distributions, the sum of the probabilities for all values which the variate may assume is unity. Thus

$$P(X \geqslant 3) = 1 - P(X = 0) - P(X = 1) - P(X = 2)$$
$$= 1 - 0{\cdot}001 - 0{\cdot}010 - 0{\cdot}044$$
$$= 0{\cdot}945.$$

Example 4.3 If a roan bull is mated with a roan cow, the probability that a single calf is red is $0{\cdot}25$. In a set of ten calves from matings of this type, what are the probabilities that the number of red calves will be $0, 1, 2, \ldots, 9, 10$?

Let X be the number of red calves. Then

$$P(X = r) = \binom{n}{r} \pi^r (1 - \pi)^{n-r}$$

$$= \binom{10}{r} (0.25)^r (0.75)^{10-r}$$

$$= \binom{10}{r} \frac{3^{10-r}}{4^{10}}.$$

r	$P(X = r)$	r	$P(X = r)$
0	$\binom{10}{0} 3^{10}/4^{10} = 0.056$	6	$\binom{10}{6} 3^4/4^{10} = 0.016$
1	$\binom{10}{1} 3^9/4^{10} = 0.188$	7	$\binom{10}{7} 3^3/4^{10} = 0.003$
2	$\binom{10}{2} 3^8/4^{10} = 0.282$	8	$\binom{10}{8} 3^2/4^{10} = 0.001$
3	$\binom{10}{3} 3^7/4^{10} = 0.250$	9	$\binom{10}{9} 3^1/4^{10} = 0.000$
4	$\binom{10}{4} 3^6/4^{10} = 0.146$	10	$\binom{10}{10} 3^0/4^{10} = 0.000$
5	$\binom{10}{5} 3^5/4^{10} = 0.058$		

It will be observed that the distribution in the table above is not symmetrical. The mode of the distribution has moved from $X = 5$ (the case when $\pi = 0.5$) to $X = 2$ for this distribution in which $\pi = 0.25$. The smaller π becomes the smaller the modal value of X and, for fixed n, the more skew the distribution becomes.

Example 4.4 For the above example, what is the probability that
(i) $X \leqslant 2$, (ii) $X \leqslant 4$, (iii) $3 \leqslant X \leqslant 6$?

(i) $\quad P(X \leqslant 2) = P(X = 0) + P(X = 1) + P(X = 2)$
$\qquad\qquad\quad = 0.056 + 0.188 + 0.282$
$\qquad\qquad\quad = 0.526.$

(ii) $P(X \leqslant 4) = P(X = 0) + P(X = 1) + P(X = 2)$
$$+ P(X = 3) + P(X = 4)$$
$$= 0.056 + 0.188 + 0.282 + 0.250 + 0.146$$
$$= 0.922.$$

(iii) $P(3 \leqslant X \leqslant 6) = P(X = 3) + P(X = 4) + P(X = 5) + P(X = 6)$
$$= 0.250 + 0.146 + 0.058 + 0.016$$
$$= 0.470.$$

4.2   Mean and Variance of the Binomial Distribution

The expected value of the variate in a binomial distribution is $n\pi$ and its variance $n\pi\delta$. This mean and variance may be calculated by either of the following two methods.

Method (i)   The expected value of the number of successes in one trial is π, since the variate assumes the values 1 and 0 with probabilities π and δ respectively, i.e.

$$E(X) = \sum p_i X_i = \pi \times 1 + \delta \times 0 = \pi.$$

Since the number of successes in n trials is the sum of the number of successes in the individual trials, it follows from an extension of

$$E(X + Y) = E(X) + E(Y)$$

that the expected value of the number of successes in n trials is $n\pi$.

Similarly, to find the variance of the distribution, the expected value of X^2 in one trial is first found.

$$E(X^2) = \sum p_i X_i^2 = \pi \times 1^2 + \delta \times 0^2 = \pi.$$

Hence the variance of the number of successes in one trial is

$$\text{var}(X) = E(X^2) - [E(X)]^2$$
$$= \pi - \pi^2$$
$$= \pi(1 - \pi)$$
$$= \pi\delta.$$

Since the number of successes in n trials is the sum of the numbers in the individual trials, and these trials are independent, the variance of the total number of successes is the sum of the variances for the separate trials and is therefore $n\pi\delta$.

Method (*ii*) The mean and variance may also be found by direct application of the formulae in the previous chapter. Thus the mean is obtained as follows. If X_i are $(n+1)$ discrete values of the variate X,

$$E(X) = \sum p_i X_i$$

$$= \delta^n . 0 + n\delta^{n-1}\pi . 1 + \binom{n}{2}\delta^{n-2}\pi^2 . 2 + \binom{n}{3}\delta^{n-3}\pi^3 . 3$$

$$+ \ldots + \pi^n . n$$

$$= n\pi\left[\delta^{n-1} + (n-1)\delta^{n-2}\pi + \binom{n-1}{2}\delta^{n-3}\pi^2 + \ldots + \pi^{n-1}\right]$$

$$= n\pi(\delta + \pi)^{n-1}$$

$$= n\pi, \qquad \text{since } \delta + \pi = 1, \text{ and hence } (\delta + \pi)^{n-1} = 1.$$

Thus the mean of the distribution is $n\pi$.

The variance may be obtained by similar, but more lengthy algebra. First $E(X^2) = \sum p_i X_i^2$ may be shown to be equal to

$$n\pi[1 + (n-1)\pi]$$

whence on subtracting $[E(X)]^2$ the variance is found to be $n\pi\delta$.

Thus for a series of n independent trials where the probability of success at each trial remains constant at π, the mean of the number of successes (X) is $n\pi$ and the variance of the number is $n\pi\delta$.

Frequently an experimenter will record not the number of successes in the series of n trials, but the proportion of successes (e.g. the proportion of diseased plants in a random sample of one hundred plants). The proportion is obtained by dividing the number observed by the number (n) of trials. Thus the expected value of the proportion is

$$n\pi/n = \pi$$

while the variance is

$$n\pi\delta/n^2 = \pi\delta/n.$$

(The division by n^2 arises from the formula

$$\text{var}(cX) = c^2 \text{ var}(X)$$

and here $c = 1/n$.)

Often the standard deviation, which is the square root of the variance, is quoted instead of the variance. The standard deviation of the number of successes is $\sqrt{(n\pi\delta)}$ while the standard deviation of the proportion of successes is $\sqrt{(\pi\delta/n)}$.

4.3 The Poisson Distribution

Another important discrete distribution is that associated with the French mathematician Poisson.

The importance of the Poisson distribution in biological research was first made apparent in connection with the accuracy of counting with a haemacytometer. If the technique of the counting process is perfect, the number of cells on each square should be theoretically distributed in a Poisson distribution.

The distribution may also be obtained in studying, say, the incidence of scurvy in a large city such as Sydney, the occurrence of noxious weed seeds in large lots of seed submitted for certification as registered or pure seed, or the number of wireworms found under a square foot of ground.

To be distributed as a Poisson variate, the variable must have two properties. Its mean must be small relative to the maximum possible number of events (theoretically, this maximum number is infinity). Also the occurrence of the event must be independent of other events. Thus, relative to the binomial distribution, the Poisson distribution applies when n is large and π is small.

Consider an agronomist studying the distribution of number of wild oat seedlings per square foot quadrat in an area treated by a herbicide to prevent emergence of wild oat seedlings. Counts of $0, 1, 2, 3, \ldots$ wild oat seedlings might be obtained. Within the area marked by the quadrat there are n 'points' where a wild oat plant might occur; n will be large but unspecified. Further, if the herbicide is preventing the emergence of the wild oat, the probability of finding a wild oat seedling at a given point in the quadrat will be very small. Here the distribution of number of seedlings would be expected to follow a Poisson distribution.

The probability function for the Poisson distribution is

$$P(X = r) = \frac{\lambda^r e^{-\lambda}}{r!}$$

where e is the base of the Naperian system of logarithms and where

$$e^{-\lambda} = 1 - \lambda + \frac{\lambda^2}{2!} - \frac{\lambda^3}{3!} + \frac{\lambda^4}{4!} - \cdots.$$

In practice, the value of $e^{-\lambda}$ is never found from this formula. Tables exist which give the values of $e^{-\lambda}$ for various values of λ.

The variable X in the Poisson distribution is discrete but may take any of the values in the infinite series

$$0, 1, 2, 3, 4, \ldots, r, \ldots.$$

The probabilities with which X takes these values are

$$e^{-\lambda}, \lambda e^{-\lambda}, \frac{\lambda^2 e^{-\lambda}}{2!}, \frac{\lambda^3 e^{-\lambda}}{3!}, \frac{\lambda^4 e^{-\lambda}}{4!}, \ldots, \frac{\lambda^r e^{-\lambda}}{r!}, \ldots.$$

The parameter in this distribution is λ. For different values of λ, different members of the family of distributions are obtained. This parameter λ is both the mean and the variance of the distribution.

4.4 Mean and Variance of the Poisson Distribution

The mean of the Poisson distribution is obtained as follows:

$$E(X) = \sum_{i=1}^{\infty} p_i X_i$$

$$= e^{-\lambda} \cdot 0 + \lambda e^{-\lambda} \cdot 1 + \frac{\lambda^2 \cdot e^{-\lambda}}{2!} \cdot 2 + \frac{\lambda^3 \cdot e^{-\lambda}}{3!} \cdot 3 + \frac{\lambda^4 \cdot e^{-\lambda}}{4!} \cdot 4 + \cdots$$

$$= \lambda e^{-\lambda} \left[1 + \lambda + \frac{\lambda^2}{2!} + \frac{\lambda^3}{3!} + \cdots \right]$$

$$= \lambda e^{-\lambda} e^{\lambda} = \lambda.$$

Again,

$$\mathrm{var}(X) = E(X^2) - [E(X)]^2.$$

By writing $E(X^2)$ as $\sum p_i X_i^2$ and substituting into the summation it can be shown, with a reasonable amount of algebra, that $E(X^2)$ is $\lambda(1+\lambda)$. Therefore,

$$\text{var}(X) = \lambda(1+\lambda)-\lambda^2 = \lambda.$$

Thus the mean and the variance of the Poisson distribution are both λ.

The relationship between the variance and the mean is one of the useful properties of this distribution, since it gives rise to a rapid test of whether an observed variable is distributed as a Poisson variate. The *coefficient of dispersion* (sample variance/sample mean) should be close to unity for variables following a Poisson distribution.

Because the Poisson distribution holds when the events are independent, it is frequently used to test for spatial randomness in nature. If the spatial distribution is uniform, as in a field where plants are sown at points on a rectangular grid, the coefficient of dispersion will be less than one. However, if the occurrence of an individual at a particular point increases the probability that other individuals will occur nearby, then clumping occurs and we have what is known as a contagious distribution. For contagious distributions the coefficient of dispersion is greater than unity. Illustrations for such distributions are given in Figure 4.1.

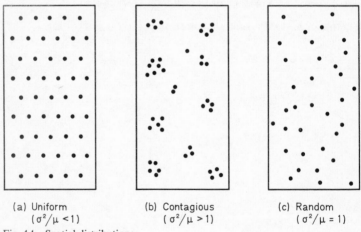

(a) Uniform   (b) Contagious   (c) Random
$(\sigma^2/\mu < 1)$ $(\sigma^2/\mu > 1)$ $(\sigma^2/\mu = 1)$

Fig. 4.1   Spatial distributions

A number of probability distributions have been considered as statistical models for cases of contagious distributions. However, only one is mentioned here—the *negative binomial distribution.* An account of the fitting of the negative binomial distribution to biological data is given by Bliss and Fisher (1953) while an interesting example of the use of this distribution is given by Baxter and Blake (1967), who considered the invasion of pieces of wheat roots by nematodes. They found that the observed number of nematodes fitted a negative binomial better than a Poisson, and concluded that each piece of root did not have an equal chance of being invaded.

4.5   Relationship of Poisson and Binomial Distributions

While the Poisson distribution is a distribution in its own right, it is closely related to the binomial. Since the Poisson distribution is appropriate for problems where the number of trials is large and the probability is small, the probability function for the Poisson may be obtained from the binomial. It can be shown that under the conditions applicable for the Poisson that

$$\lim_{n \to \infty} \binom{n}{r} \pi^n (1-\pi)^{n-r} = \frac{\lambda^r e^{-\lambda}}{r!}$$

where λ has been written for $n\pi$.

Thus the Poisson is a limiting form of the binomial.

The mean and variance may be deduced from those of the binomial by putting $\pi = \lambda/n$ and letting n tend to infinity. Thus

$$\text{Mean} = E(X) = \lim_{n \to \infty} n\pi$$

$$= \lim_{n \to \infty} n\lambda/n = \lambda \, .$$

$$\text{Variance} = \lim_{n \to \infty} n\pi(1-\pi)$$

$$= \lim_{n \to \infty} n\lambda/n \left[1 - \frac{\lambda}{n} \right] = \lambda \, .$$

The agreement between the terms in the binomial and Poisson distribution may be appreciated by considering the two probability functions

$$P(X = r) = \binom{n}{r}\pi^r\delta^{n-r}, \qquad r = 0, 1, 2, \ldots, n,$$

and
$$P(X = r) = \frac{e^{-\lambda}\lambda^r}{r!}, \qquad r = 0, 1, 2, \ldots.$$

for $n = 10$, $\pi = 0.1$ and $\lambda = 1.0$ (Case 1),
and $n = 100$, $\pi = 0.01$ and $\lambda = 1.0$ (Case 2).

Table 4.1 gives the probabilities for the two distributions. The agreement is seen to be not too bad even when $n = 10$ and is very good for $n = 100$.

TABLE 4.1 *Comparison of probabilities for binomial and Poisson distribution*

	Case 1		Case 2	
r	Binomial	Poisson	Binomial	Poisson
0	0·3487	0·3679	0·3660	0·3679
1	0·3874	0·3679	0·3697	0·3679
2	0·1937	0·1839	0·1849	0·1839
3	0·0574	0·0613	0·0610	0·0613
4	0·0112	0·0153	0·0149	0·0153
5	0·0015	0·0031	0·0029	0·0031
6	0·0001	0·0005	0·0005	0·0005
7	0·0000	0·0001	0·0001	0·0001

Example 4.5 The mortality rate for a certain disease is 6 per 1000. What is the probability of exactly 4 deaths from this disease in a randomly selected group of 500?

From the binomial distribution, the probabilities of various numbers of deaths, $0, 1, 2, \ldots, 500$ would be given by the terms of the expansion

$$(\delta + \pi)^{500}$$

where $\delta = 1 - \pi$ and $\pi = 0.006$.
Thus the probability of 4 deaths is

$$\binom{500}{4}(0.006)^4(0.994)^{496}.$$

The difficulty in computing such a quantity is immediately apparent. However, this problem differs from the type of problem to which the binomial distribution is usually applied. First, the number in the group is very large and second, the probability is very small. These are the conditions under which the Poisson distribution is applicable, and the probabilities in the binomial distribution

$$\delta^n, \binom{n}{1}\delta^{n-1}\pi, \binom{n}{2}\delta^{n-2}\pi^2, \ldots, \binom{n}{r}\delta^{n-r}\pi^r, \ldots, \binom{n}{n}\pi^n$$

are approximated very closely by the terms of the Poisson distribution

$$e^{-\lambda}, \lambda e^{-\lambda}, \lambda^2 e^{-\lambda}/2!, \ldots, \lambda^r e^{-\lambda}/r!, \ldots, \lambda^n e^{-\lambda}/n!, \ldots.$$

Thus

$$P(X = 4) = \lambda^4 e^{-\lambda}/4!$$

where $\lambda = n\pi = 500 \times 0{\cdot}006 = 3$. Hence

$$P(X = 4) = 3^4 e^{-3}/4!$$
$$= 3^4 \times 0{\cdot}0498/4!$$
$$= 0{\cdot}168.$$

Example 4.6 Suppose that a truck-load of wheat is being sampled prior to certification and that the sample chosen at random consists of one half-pound or 8000 seeds. Each sample is inspected in detail for the presence of weed seeds. Assuming that in the population sampled the weed seeds are randomly distributed and that there is an average of four weed seeds in every pound, what is the probability that the sample chosen will have two weed seeds?

This probability could be obtained using the binomial distribution. It is

$$P(X = 2) = \binom{8000}{2}\pi^2\delta^{7998}$$

where $\pi = 4/16\,000$ and $\delta = 1 - \pi$. However, since n is very large and π is very small, an excellent approximation can be obtained from the Poisson distribution. In this case,

$$P(X = 2) = \lambda^2 e^{-\lambda}/2!,$$

$$\lambda = n\pi = 8000 \times 4/16\,000 = 2.$$

Hence

$$P(X = 2) = 2^2 e^{-2}/2!$$
$$= 2 \times 0.135$$
$$= 0.270.$$

Exercises

4.1   For the preceding example, what is the probability that the number of weed seeds will be three or more?

4.2   Suppose that there are 5000 heads of wheat per plot in a series of 400 plots. On examination of the plots it is noted that a small proportion of heads are affected by a particular disease. In order to study the hypothesis of independence of disease from head to head, it is proposed to count the number of diseased heads in each plot. If it were known that in this series of plots, the mean number of diseased heads were 6 per 10 000, what would be the expected number of plots having 2 diseased heads?

4.3   Assume that when *Toxoptera aurantii* (the black citrus aphid) are sprayed with a concentration of 0.02% malathion, four-tenths of the insects die. Find the mean and standard deviation for the number dying when samples of one hundred of the same insects are sprayed (use the binomial distribution).

4.4   For the previous exercise, what is the standard deviation if it were assumed that the distribution of numbers dead were a Poisson distribution?

4.5   A simple experiment is designed to test the efficiency of an insecticide in killing mosquitoes. If it is assumed that, on the average, this insecticide kills 80% of the mosquitoes, what is the probability that, after the experiment
  (i) there will be 85 dead in a sample of 100;
 (ii) 75 dead in a sample of 100?

4.6   In a wheat field of approximately 2 acres, the distribution of wild oat plants was studied from counts made on 400 randomly-chosen quadrats. The following frequency distribution of counts per quadrat was obtained. Calculate the coefficient of dispersion to examine the hypothesis that individual plants are occurring randomly in the field.

Number of plants per quadrat (X)	0	1	2	3	4	5
Observed frequency (f)	210	130	46	11	2	1

REFERENCES

BAXTER. R. I. and BLAKE, C. D. (1967). 'Invasion of Wheat Roots by *Pratylenchus thornei*', *Nature, 215*, 5106, 1168–1169.

BLISS, C. I. and FISHER, R. A. (1953). 'Fitting the Negative Binomial Distribution to Biological Data and Note on the Efficient Fitting of the Negative Binomial', *Biometrics*, 9, 176–200.

COLLATERAL READING

BLISS, C. I. (1967). *Statistics in Biology*. McGraw-Hill, New York. Chapter 2.

FINNEY, D. J. (1964). *An Introduction to Statistical Science in Agriculture* (2nd edition). Oliver & Boyd, Edinburgh; John Wiley, New York. Chapter 3.

GOODMAN, R. (1957). *Teach Yourself Statistics*, English Universities Press, London. Chapters 3 & 4.

GOULDEN, C. H. (1952). *Methods of Statistical Analysis* (2nd edition). John Wiley, New York. Chapter 3.

SNEDECOR, G. W. (1956). *Statistical Methods* (5th edition). The Iowa State University Press, Ames, Iowa. Chapter 16.

STEEL, R. G. D. and TORRIE, J. H. (1960). *Principles and Procedures of Stattistics*. McGraw-Hill, New York. Chapter 20.

CHAPTER 5

The Normal and χ^2 Distributions

5.1 Introduction

There are a large number of different distributions which are useful in biometry. In the previous chapter an account was given of two distributions for which the variate was discrete. In this chapter a study is commenced of one of the most commonly used distributions where the variable is continuous. This is the *normal* distribution.

The term normal does not imply that other distributions are abnormal. The name 'normal' is one of the names given to the distribution which is considered in the first part of this chapter. A comprehensive account of the binomial distribution is contained in a posthumous work by James Bernoulli, published twenty years before De Moivre in 1733 first presented the equation of the normal distribution. This was later rediscovered and developed by Gauss (1809) and by Laplace (1812). For these reasons the normal distribution is also called the De Moivre, Gaussian or Laplace distribution.

The normal distribution is an extremely important distribution since it forms the basis for many other distributions which are used in statistics. Later in this chapter its relationship to the χ^2 distribution will be considered. Then at a later stage in the development of the subject of biometry it will be seen that the ratio of two χ^2 variates follows an F distribution. The F distribution is used in the analysis of variance, probably the most common statistical technique used by agricultural and biological scientists.

The normal distribution is important not only because it forms the basis for other distributions, but because it is a mathematical idealization of the actual distribution of innumerable continuous variables which biologists have to deal with in practice.

The normal distribution may be derived from elementary assumptions in a number of ways with a knowledge of mathematics above that which is assumed for the reader of this text. The two rather intuitive approaches which are used here would not satisfy a mathematician, but they are presented because they have been found to have heuristic value.

The first approach depends on the fact that the normal distribution is related to the binomial. Now if we consider a continuous variable such as plant height, it will be realized that many factors, environmental and genetic, contribute to the final height of the plant. Suppose as a simplification of the way in which these factors determine the final height

(i) that each of the factors exists in only two states—present or absent;

(ii) that if the factor is present it contributes one unit to the final height;

(iii) that each of the factors has the same effect on final yield and that these effects are additive; thus if five out of seven factors are present the plant has five units of height;

(iv) for each factor there is an equal probability of its being present or absent.

With these assumptions, consider now the binomial distribution with parameters n and π. If only one factor is operating, then $n = 1$, $\pi = \frac{1}{2}$ and the two height classes (T and t, say) are obtained with equal probability. In a population of plants, half would be T and have one unit of height, the other half would be t and have zero units of height. If the number of factors were two, then the height classes would be TT (2 units), Tt (1 unit), tt (0 units) and these would occur with probabilities 0·25, 0·5 and 0·25 respectively. If ten factors were influencing height, there would be eleven height-classes, and these would occur with the probabilities presented in Example 4.1.

The eleven probabilities are found to agree very well with the areas under the curve

$$y = \frac{1}{\sigma\sqrt{(2\pi)}} \exp\left[\frac{-(X-\mu)^2}{2\sigma^2}\right]^*$$

with $\mu = n\pi = 10 \times 0.5$, $\sigma = \sqrt{(n\pi\delta)} = \sqrt{(10 \times 0.25)}$, and above the intervals

$$-\tfrac{1}{2} \leqslant X \leqslant \tfrac{1}{2}, \quad \tfrac{1}{2} \leqslant X \leqslant 1\tfrac{1}{2}, \quad 1\tfrac{1}{2} \leqslant X \leqslant 2\tfrac{1}{2}, \dots, 9\tfrac{1}{2} \leqslant X \leqslant 10\tfrac{1}{2}.$$

* The notation exp (x) means e^x, e being the base of natural logarithms.

Now it can be shown mathematically that provided neither π nor δ is of the order of $1/n$ or less, as n becomes larger and larger the binomial probability distribution approaches more and more nearly a curve whose equation is

$$y = \frac{1}{\sigma\sqrt{(2\pi)}} \exp\left[\frac{-(X-\mu)^2}{2\sigma^2}\right]. \qquad (5.1)$$

This limit which is approached by the binomial distribution is called the normal distribution. The family of curves specified by equation (5.1) is characterized by two parameters, μ and σ, and

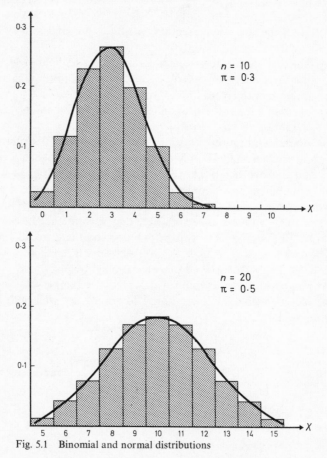

Fig. 5.1 Binomial and normal distributions

does not contain either n or π. The relationship between the binomial and normal distribution is apparent from an examination of Figure 5.1. Here a normal curve is superimposed on the histograms (rather than bar diagrams as they should be) of two binomial distributions ($n = 10, \pi = 0\cdot3; n = 20, \pi = 0\cdot5$).

Now it may be thought that the initial assumptions were rather stringent. It is interesting to note that they may be considerably relaxed. Even when the two states of a factor are not present with equal probability, i.e. if $\pi \neq \delta$ (or $\pi \neq \frac{1}{2}$), the binomial distribution still approaches normality as n is indefinitely increased. Again, if the factor exists in more than two states the normal distribution is still the limit as n is increased. Further, the different factors may occur with different probabilities and have different quantitative effects and normality is still approached as n is increased to infinity, provided the effects are additive and independent. Hence it is not surprising that many of the quantitative variables which the biologist has to study, are found to have normal (or at least very nearly normal) distributions.

The preceding introduction to the normal distribution might be considered a somewhat mathematical one. An alternative approach is the following.

Suppose that a histogram is drawn for the relative frequency distribution of heights for a random sample of 300 wheat plants. Since the areas of the rectangles in the histogram are equivalent to relative frequencies, the sum of the areas of all the rectangles is unity.

If now it were possible to measure the heights with much greater accuracy and if instead of 300 plants there were 3000, a slightly different histogram of relative frequencies would be drawn. While the same base line would be used, the class intervals would be smaller. The rectangles would be narrower than for the smaller sample and there would be more of them. Again, though, the areas would sum to unity.

Imagine now the sample size being increased still further. When there were 300 000 heights, the class intervals would be very much smaller and there would be many more rectangles, each much narrower than when the sample size was 3000. If now a polygon were superimposed on this third histogram, the polygon would give the appearance of an almost smooth curve.

The larger the sample becomes, the closer the agreement between the polygon and the continuous curve. This idea is illustrated by the relative frequency histograms presented in Figure 5.2.

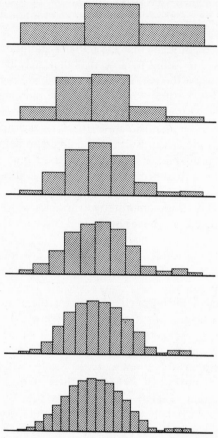

Fig. 5.2 A series of relative frequency histograms to illustrate approach to a continuous curve

Also as the sample size was increased the sample mean ($\bar{x}$) would tend to approach a constant (μ) and the sample variance (s^2) would tend to a constant (σ^2). The frequency polygon would approach a continuous curve whose equation would be

$$y = \frac{1}{\sigma\sqrt{(2\pi)}}\exp\left[\frac{-(X-\mu)^2}{2\sigma^2}\right].$$

Just as the sum of the areas of the rectangles in the relative frequency histogram is unity, the area under this curve between $-\infty$ and ∞ is unity. In other words

$$\frac{1}{\sigma\sqrt{(2\pi)}} \int_{-\infty}^{\infty} \exp\left[\frac{-(X-\mu)^2}{2\sigma^2}\right] dX = 1.$$

5.2 Definition and Properties of the Normal Distribution

For a continuous variate it is useless to speak of the probability of any particular value. Instead of this, the probability that the variate will fall within a specified interval is considered. For continuous variates, the probability of the variate falling within the infinitesimal interval $X-\frac{1}{2}dX$ to $X+\frac{1}{2}dX$ is considered. This is expressible in the form $\phi(X)\,dX$ where $\phi(X)$ is a continuous function of X called the *probability density* or the *probability function* of X.

A continuous variate X is normally distributed, with mean μ and standard deviation σ when the range of the variate is from $-\infty$ to ∞ and the probability density is given by

$$\phi(X) = \frac{1}{\sigma\sqrt{(2\pi)}} \exp\left[\frac{-(X-\mu)^2}{2\sigma^2}\right].$$

One of the chief properties of the equation for the probability function of a normal curve is that it can be written to depend only on the mean (μ) and the standard deviation (σ). The only difference between the equations for any two normal distributions is that they have different values of one or both of these quantities. (Numerical values that serve in this way to identify a particular distribution curve, as a member of a whole family of curves, are called parameters.)

The probability curve, $y = \phi(X)$, is symmetrical about the line $X = \mu$ (a line through the mean of the distribution). This can be seen by putting $X = \mu + a$ and $X = \mu - a$ in the equation for the probability curve. For both of these values of X,

$$y = \frac{1}{\sigma\sqrt{(2\pi)}} \exp\left[\frac{-a^2}{2\sigma^2}\right].$$

In geometrical terms the normal curve is bell-shaped, symmetric, and asymptotic to the horizontal axis. Further, the mean, median and mode coincide.

The parameter μ determines the position of the curve, while the standard deviation σ indicates the spread of the distribution. Figure 5.3 shows the effect of change in the two parameters. In the

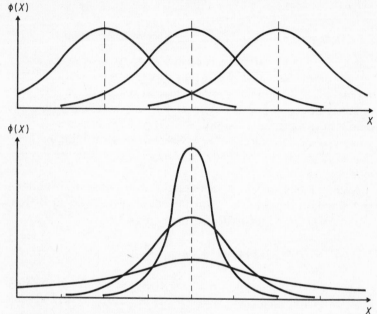

Fig. 5.3   Different normal distributions

top half of the figure, normal curves with different means μ and the same standard deviation σ are presented, while the lower half shows normal curves with the same mean but different standard deviations.

Exercise 5.1   On a single sheet of one-tenth by one-tenth inch graph paper, draw the probability curves for the following normal distributions over the range of X indicated. Note the change in position with change in μ and the change in spread with change in σ.

    (i) $\mu = 0, \sigma = 1$; $(-2\cdot5 \leqslant X \leqslant 2\cdot5)$.
    (ii) $\mu = 2, \sigma = 1$; $(-0\cdot5 \leqslant X \leqslant 4\cdot5)$.
    (iii) $\mu = 0, \sigma = 3$; $(-7\cdot5 \leqslant X \leqslant 7\cdot5)$.
    (iv) $\mu = 2, \sigma = 3$; $(-5\cdot5 \leqslant X \leqslant 9\cdot5)$.

What is the area under each curve (to the nearest 0.1 in^2)?

In drawing the curves it is suggested that the value of y be obtained for the following points on the X axis:

(i) $X = 0, 0.5, 1.0, 1.5, 2.0, 2.5$ and hence $-0.5, -1.0, -1.5, -2.0, -2.5$.

(ii) $X = 2, 2.5, 3.0, 3.5, 4.0, 4.5$ and hence $1.5, 1.0, 0.5, 0, -0.5$.

(Note that the ordinates in (i) and (ii) are the same.)

(iii) $X = 0, 1.5, 3.0, 4.5, 6.0, 7.5; -1.5, -3.0, -4.5, -6.0, -7.5$.

(iv) $X = 2, 3.5, 5.0, 6.5, 8.0, 9.5; 0.5, -1.0, -2.5, -4.0, -5.5$.

5.3   Probabilities and Relative Frequencies for Various Intervals

For a normal variate z with mean $\mu = 0$ and standard deviation $\sigma = 1$, the probability density is

$$\phi(z) = \frac{1}{\sqrt{(2\pi)}} e^{-\frac{1}{2}z^2}. \tag{5.2}$$

Such a variate is generally referred to as a *standardized normal variate*.

The area under the curve (5.2) between $z = -\infty$ and $z = \infty$ is unity, and the probability corresponding to any interval in the range of the variate is represented by the area under the probability curve and above the interval. Since the total area under the curve from $-\infty$ to ∞ is unity and since the normal curve is symmetrical, the area from $-\infty$ to zero is one-half.

In Table I (Appendix) the areas under the curve from $z = 0$ to various positive z values are presented. Because of the symmetry of the curve the areas presented in Table I can be used to find areas associated with all intervals irrespective of whether the end points of the intervals are positive or negative.

A study of Table I shows that the area from $z = 0$ (the mean) to $z = 1$ (the standard deviation) is 0.3413. Hence the area under the probability curve for a standardized normal variate between $z = -1$ and $z = +1$ is 0.6826 (from the symmetry of the curve). Thus the probability that a random value of a standardized normal variate will lie between -1 and $+1$ is given by

$$P(-1 \leqslant z \leqslant +1) = 0.6826.$$

The area outside this interval is 0·3174, which is less than one-third. In other words, the probability that a random value of a standardized normal variate will deviate by more than unity from its mean is less than one-third.

For $z = 1·96$, the area between zero and 1·96 is 0·4750 and the area for the interval $-1·96 \leqslant z \leqslant 1·96$ is 0·95. The area outside this interval is therefore 0·05, i.e.

$$P((z \leqslant -1·96) \text{ or } (z \geqslant 1·96)) = P(z \leqslant -1·96) + P(z \geqslant 1·96)$$
$$= 0·025 + 0·025$$
$$= 0·05.$$

Thus, the probability that a random value of z will deviate more than 1·96 from the mean is 0·05.

Similarly the area outside the interval $z = -2·58$ to $z = 2·58$ is 0·01 or 1%. Deviations in excess of four times the standard deviation are extremely rare for normal variates and will seldom, if ever, be encountered in small samples.

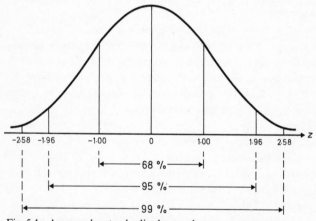

Fig. 5.4   Areas under standardized normal curve

The above probabilities, which are illustrated in Figure 5.4, have been found for a standardized normal variate. However, in practice biologists have to deal with variates which have means different from zero and standard deviations different from unity. In the case of yields (X) from a new strain, the plant breeder might wish to obtain the probability that his new strain would give a yield greater than some specified quantity. In this case the mean μ would be different from zero and the standard deviation σ would be different from unity. Fortunately, the probabilities in cases such as these can be found from those for a standardized normal variate by using the transformation

$$z = (X - \mu)/\sigma. \tag{5.3}$$

The relative position, size and proportions of the curve (5.1) depend on the values of μ and σ, as these parameters determine the origin and scale of measurement. The effect of the transformation (5.3) is to shift the mean μ of the X distribution to zero on the horizontal scale. It also results in a stretching or compression of the horizontal scale by a factor $1/\sigma$. However, this change in the horizontal scale is compensated for by a change in the vertical scale by a factor σ. The total area under either the z or X curves between $-\infty$ and ∞ is unity. It can also be shown that the area under the curve (5.1) between two points X_1 and X_2 is equal to the area under the curve (5.2) between the two corresponding points z_1 and z_2.

Hence by use of the transformation (5.3) and the result concerning the equality of areas, the probability that a random value of a normal variate with mean μ and a variance σ^2 will lie between two values X_1 and X_2 is obtained as follows:

$$P(X_1 \leqslant X \leqslant X_2) = P(z_1 \leqslant z \leqslant z_2)$$

where z is a standardized normal variate and $z_i = (X_i - \mu)/\sigma$.

Thus only one set of tables—those for z—is necessary to solve all problems of probabilities for normal distributions.

Example 5.1 The leaf area per plant (X) for a particular wheat variety is normally distributed with mean 100 cm^2 and standard deviation 10 cm^2. What is the probability that a plant chosen at random will have a leaf area (X) such that

(i) $80.4 \leqslant X \leqslant 119.6$;

(ii) $-\infty \leqslant X \leqslant 100$;

(iii) $125.8 \leqslant X \leqslant \infty$.

The transformation $z = (X - 100)/10$ will be used in each case to transform to a z variate and hence to obtain the required probability.

(i) The required probability is given by

$$P(80\cdot4 \leqslant X \leqslant 119\cdot6) = P(z_1 \leqslant z \leqslant z_2).$$

Here

$$z_1 = {}_1(X_1 - \mu)/\sigma = (80\cdot4 - 100)/10 = -1\cdot96,$$
$$z_2 = (X_2 - \mu)/\sigma = (119\cdot6 - 100)/10 = 1\cdot96,$$

which gives

$$P(-1\cdot96 \leqslant z \leqslant 1\cdot96) = P(-1\cdot96 \leqslant z \leqslant 0) + P(0 \leqslant z \leqslant 1\cdot96)$$
$$= 0\cdot475 + 0\cdot475 = 0\cdot95.$$

Thus the probability of randomly choosing a plant with a leaf area between $80\cdot4$ cm^2 and $119\cdot6$ cm^2 is 95%. The proportion of plants having a leaf area outside this interval is very small for a variety for which leaf area is normally distributed with a mean of 100 cm^2 and a standard deviation of 10 cm^2.

It should be noted that $80\cdot4$ is $1\cdot96$ standard deviations below the mean of 100, while $119\cdot6$ is $1\cdot96$ standard deviations above the mean.

(ii) The required probability is given by

$$P(-\infty \leqslant X \leqslant 100) = P(z_1 \leqslant z \leqslant z_2).$$

Here

$$z_1 = (X_1 - \mu)/\sigma = (-\infty - 100)/10 = -\infty$$

$$z_2 = (X_2 - \mu)/\sigma = (100 - 100)/10 = 0,$$

which gives

$$P(-\infty \leqslant z \leqslant 0) = 0\cdot5.$$

Thus the probability that a plant chosen at random will have a leaf area less than the mean is $0\cdot5$. This is a property of the normal distribution and is a consequence of its symmetry about the mean.

(iii) The required probability is given by

$$P(125\cdot8 \leqslant X \leqslant \infty) = P(z_1 \leqslant z \leqslant z_2).$$

Here $z_1 = 2\cdot58$ and $z_2 = \infty$, which gives

$$P(2\cdot58 \leqslant z \leqslant \infty) = 0\cdot005.$$

The probability of a plant having a leaf area greater than $125\cdot8$ cm^2 is very small. While the normal distribution theoretically has a maximum value for the variable of ∞, it is seen here that the probability of obtaining a value many standard deviations from the mean is very small.

Example 5.2 In the previous example, Table I was used to obtain a probability for a given value of z. Frequently this table has to be used in the reverse order; given a probability, what is the corresponding value of z?

Suppose that for the population of wheat plants in Example 5.1 the plant heights (Y) are normally distributed with mean 100 cm and variance 16 cm². Find Y_0 such that

(i) $P(Y \geqslant Y_0) = 0.01$; (ii) $P(Y \geqslant Y_0) = 0.90$

(i) Now
$$0.01 = P[Y \geqslant Y_0]$$
$$= P[\{(Y-\mu)/\sigma\} \geqslant \{(Y_0 - 100)/4\}]$$
$$= P[z \geqslant \{(Y_0 - 100)/4\}].$$

The z table gives $P[z \geqslant 2.32] = 0.01$. Hence
$$(Y_0 - 100)/4 = 2.32$$
$$Y_0 = 109.28.$$

(ii) Again
$$0.90 = P[Y \geqslant Y_0]$$
$$= P[z \geqslant (Y_0 - 100)/4].$$

The z table gives $P[z \geqslant -1.28] = 0.90$. Hence
$$(Y_0 - 100)/4 = -1.28,$$
$$Y_0 = 94.88.$$

Exercises

5.2 (a) Given that the weights of calves (W) are normally distributed with mean 10 kg and standard deviation 4 kg, find $P(W \leqslant 18)$, $P(2 \leqslant W \leqslant 14)$. What is the probability that a calf will weigh more than 10 kg; more than 18 kg?

(b) Given a normal distribution (X) with $\mu = 50$, $\sigma = 2$, find $P(X \geqslant 53.92)$, $P(46.08 \leqslant X \leqslant 52)$. Illustrate on two graphs of the normal distribution, the areas corresponding to the two probabilities.

5.3 (a) Assuming that the differences (D) in plant heights over an arbitrary period of time are normally distributed with mean 10 cm and variance 9 cm^2, find D_1, D_2 and D_3 such that $P(D \leqslant D_1) = 0.025$, $P(D_2 \leqslant D \leqslant D_3) = 0.95$. What biological interpretation might be given to your results?

(b) Given a normal distribution of Y with $\mu = 20$ and $\sigma^2 = 25$, find Y_1 and Y_2 such that $P(Y \leqslant Y_1) = 0.25$, $P(Y \geqslant Y_2) = 0.90$. Mark Y_1 and Y_2 on a graph of the normal distribution.

5.4 The χ^2 Distribution

The χ^2 distribution is used by agricultural scientists and biologists—either explicitly or implicitly—in many different areas. It may be used to examine whether samples may be reasonably regarded as having been drawn from populations whose variances are known, and for comparing the agreement between observed and expected frequency distributions. It is also used implicitly in the analysis of variance (one of the most commonly used techniques in biometry).

As stated at the beginning of this chapter the χ^2 distribution is related to the normal distribution. For this reason it is introduced in this chapter. However, examples of its use in biometry are given in later chapters.

The chi-square variate with ν degrees of freedom is defined as the sum of squares of ν independent, normally distributed variates each with zero mean and unit variance.

Thus, if X_i is normally distributed with mean μ_i and variance σ_i^2,

$$\chi^2 = \sum_{i=1}^{\nu} \frac{(X_i - \mu_i)^2}{\sigma_i^2}.$$

The distribution of χ^2 depends on the number of independent variates, i.e. on the *degrees of freedom* (ν). The only parameter of the distribution is ν and there is a χ^2 distribution for each number (ν) of degrees of freedom. χ^2 with 1 degree of freedom is the square of a standardized normal variate.

An important property of the χ^2 distribution is that its mean value is the degrees of freedom, ν.

Some χ^2 probability curves with degrees of freedom $\geqslant 2$ are shown in Figure 5.5.

Obviously, since χ^2 involves the sum of a number of squares it cannot be negative. Thus, the range of χ^2 is from 0 to ∞. While the mode of the distribution curve lags behind the degrees of freedom (mode $= v-2$), the curves become more and more symmetrical as the number of degrees of freedom increases.

Table II in the appendix gives for various probabilities (α) the percentage points χ^2_α of the χ^2 distribution where χ^2_α is defined to be the value of χ^2 such that $P(\chi^2 \geqslant \chi^2_\alpha) = \alpha$. In order to use the table

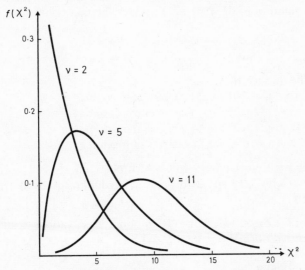

Fig. 5.5 Some χ^2 probability curves with degrees of freedom $\geqslant 2$

it is necessary to know the number of degrees of freedom (v) of the particular χ^2 distribution being investigated. The degrees of freedom (v) are given in the left-hand column of the table and values of χ^2 in the body of the table.

Example 5.3 Find the value of χ^2 with three degrees of freedom that is exceeded with probability 0.50, i.e. find χ^2_0 such that

$$P(\chi^2_{v=3} \geqslant \chi^2_0) = 0.50.$$

Enter the χ^2 table for three degrees of freedom and read under the column headed 0.50. Here $\chi^2 = 2.366$ and thus

$$P(\chi^2_{v=3} \geqslant 2.366) = 0.50,$$

i.e. $\chi^2_0 = 2.366$.

χ_0^2 is the median for this χ^2 distribution. The mean value for this distribution is three, the number of degrees of freedom. The median is less than the mean since the distribution is skewed to the right.

Example 5.4 Find the probability with which an observed χ^2 = 35·02 with twenty degrees of freedom will be exceeded.

Enter the χ^2 table for twenty degrees of freedom and read across for the number 35·02. It is found under the column headed 0·02. Hence

$$P(\chi^2 \geqslant 35·02) = 0·02.$$

Exercises

5.4 Find χ_0^2 such that

$P(\chi^2 \geqslant \chi_0^2) = 0·01$ for 20 degrees of freedom;

$P(\chi^2 \geqslant \chi_0^2) = 0·05$ for 1 degree of freedom;

$P(\chi^2 \geqslant \chi_0^2) = 0·02$ for 14 degrees of freedom;

$P(\chi^2 \geqslant \chi_0^2) = 0·95$ for 28 degrees of freedom.

5.5 Find the probabilities with which observed χ^2 of 11·07, 30·58 and 12·70 with 5, 15 and 25 degrees of freedom respectively will be exceeded.

5.6 If $(Z_1, Z_2, \ldots, Z_{10})$ are 10 random observations from the standardized normal population and if

$$S = \sum_{i=1}^{10} Z_i^2 \text{ find}$$

(a) S_0 such that $P(S \geqslant S_0) = 0·20$;

(b) $P(S \geqslant 18·307)$;

(c) $P(3·940 \leqslant S \leqslant 23·209)$.

5.7 If (X_1, X_2, X_3, X_4) are four random observations from a normal population having mean μ and variance σ^2, what is the expected value of the distribution of the following sum of squares

$$\left(\frac{X_1 - \mu}{\sigma}\right)^2 + \left(\frac{X_2 - \mu}{\sigma}\right)^2 + \left(\frac{X_3 - \mu}{\sigma}\right)^2 + \left(\frac{X_4 - \mu}{\sigma}\right)^2 ?$$

REFERENCES

GAUSS, K. F. (1809). *Werke*, **4**, Göttingen, 1–93.
LAPLACE, P. S. (1812). *Théorie Analytique des Probabilités*. Paris.

CHAPTER 6

Some Sampling Theory

6.1 Random Sampling from a Population

In order to examine a large population with respect to a particular variable, or variables, the experimenter must choose a sample of individuals from the population and, from measurements or observations made on the sample, make an estimate of the parameters of the distribution of the variable in the population. Suppose that the variable under consideration is the height of a plant; then the assemblage of heights of all the plants, either in a large specified group or an infinite hypothetical group, is called a *population* of heights, and those of the individuals in the sample are known collectively as a *sample* of heights from that population. The population need not be one of heights—it may be a population of weights of animals, yields of grain, percentage of diseased plants or the number of seeds germinating in a pot of soil.

The theory of sampling is concerned with estimating the parameters of the population from the observations made on the sample; with assessing the precision of these estimates; and with making tests of significance concerning the population parameters. Fundamental to the theory is the concept of random sampling which was defined previously.

A sample of n members is a *random sample* if it is chosen in such a manner that each possible sample of size n has the same probability of being chosen.

Associated with the concept of a random sample is that of a *simple sample*. Simple sampling is random sampling with the further provision that the probabilities of selection of members remain constant throughout the selection. Hence in sampling a finite population, the individual must be returned to the population before the next sampling if a simple sample is to be drawn.

However, a population whose distribution is continuous (e.g.

yield, height, age) contains an infinite number of values in any finite interval in the range of the variable. In drawing a finite random sample from such a population, the probability associated with any interval remains unchanged. Hence, a finite random sample from a population whose distribution is continuous is a simple sample. However, it is conventional to refer to such a sample as a 'random sample'.

6.2 Sampling Distribution: Standard Error

The distribution of a qualitative or quantitative, discrete or continuous variable in a population has certain parameters. For a binomial variate, the parameters are n and π, and for a Poisson variate the parameter is λ, while in the case of the normal distribution there are two parameters, μ and σ^2.

For each random sample, statistics such as proportions of individuals in the sample with a specified characteristic, or the sample mean, or the sample variance, etc., may be calculated. A *statistic* is a function of the observed variables. Statistics are usually calculated to obtain estimators of the parameters of the population. The value of the statistic will vary from sample to sample and the distribution of the statistic obtained by repeated continued random sampling is called the *sampling distribution of the statistic* (in some places, the derived distribution). The sampling distribution is determined by the nature of the population and the size of the sample.

The estimator of a parameter is often represented by the same Greek letter as that used to represent the parameter. However, a hat or cap is placed above the letter when it is used for an estimator. This is done to differentiate between the parameter and its estimator. For example, $\hat{\mu}, \hat{\sigma}^2, \hat{\pi}, \hat{\beta}$ represent estimators of the parameters $\mu, \sigma^2, \pi, \beta$.

If the mean of the sampling distribution of a statistic is equal to a parameter, θ, of the population from which the sample was drawn, the statistic is said to be an unbiased estimator of θ, i.e. a statistic is *unbiased* if its expectation in random sampling equals the population parameter.

The standard deviation of the sampling distribution of a statistic is called the *standard error* (s.e.) of that statistic. There may be different unbiased statistics or estimators for a particular population parameter (e.g. the sample mean $\bar{x}, \sum a_i X_i$ where the a_i are constants such that $\sum a_i = 1$, and the sample median are three unbiased estimators of the normal population mean, μ). The statistic with the smallest standard error is said to be the *most efficient*.

In addition to the criteria of unbiasedness and efficiency, two other criteria are used when evaluating different statistics as estimators of a particular population parameter. These are *consistency* and *sufficiency*.

However, the concepts of consistency and sufficiency are mathematical ones which will not be used in this text. They are mentioned to indicate that statisticians, when choosing a statistic to estimate a parameter, have certain criteria on which to assess the statistic as an estimator.

6.3   Sampling of Continuous Variables: the Linear Additive Model

The sampling of continuous variables will now be considered. As most continuous variables with which the biologist has to deal are normally distributed or may be transformed by an appropriate transformation to a normal variable, random sampling from a normal population will be considered in the following sections.

To explain the make-up of biological data, mathematical models are used a great deal. A model used frequently to represent the data of a random sample is

$$X_i = \mu + \varepsilon_i$$

where μ represents the mean of the population sampled and ε_i are the deviations of the observed values from the mean, μ. This is known as a *linear, additive model*. The model states that the ε_i are distributed with mean zero, i.e. $E(\varepsilon_i) = 0$.
Hence

$$
\begin{aligned}
E(X_i) &= E(\mu) + E(\varepsilon_i) \\
&= \mu + E(\varepsilon_i) \quad \text{since } \mu \text{ is a fixed parameter} \\
&= \mu \quad \text{since } E(\varepsilon_i) = 0.
\end{aligned}
$$

If sampling is random, the ε_i are independent, i.e. $E(\varepsilon_i \varepsilon_j) = E(\varepsilon_i)E(\varepsilon_j) = 0$ if $i \neq j$. Again, if the population (X) which is sampled is normal with variance σ^2, the deviations (ε) are normally distributed with variance σ^2. Thus

$$\begin{aligned}
\sigma^2 &= \text{variance } (\varepsilon_i) \\
&= E(\varepsilon_i^2) - [E(\varepsilon_i)]^2 \\
&= E(\varepsilon_i^2) \qquad \text{since } E(\varepsilon_i) = 0.
\end{aligned}$$

6.4 Sampling Distribution of the Mean

The distribution of the arithmetic means $(\bar{x})$ of random samples of a given size from a normal population is of great importance. Research workers in the agricultural and biological sciences are continually dealing with treatment or strain means. The manner in which such means vary from one experiment to the next must be understood by all experimentalists who wish to study and interpret their data by means of biometrical techniques. The importance to be attached to the theorem presented in this section may be appreciated when it is realized that this is the only result presented in italics in this text.

Let the sampled population (X) have mean μ and variance σ^2. Then the sampling distribution of $\bar{x}$, of random samples of size n has μ for its mean and σ^2/n for its variance. The s.e. of the mean $\bar{x}$ is $\sigma/\sqrt{n}$.

The statistical model is
$$X_i = \mu + \varepsilon_i, \qquad (i = 1, 2, \ldots, n),$$
where $E(\varepsilon_i) = 0$, $E(\varepsilon_i^2) = \sigma^2$, and $E(\varepsilon_i \varepsilon_j) = 0$ if $i \neq j$.
Then

$$\bar{x} = \mu + \sum_{i=1}^{n} \varepsilon_i/n,$$

$$E(\bar{x}) = E(\mu) + E(\sum \varepsilon_i/n)$$

and since expected value of a sum equals the sum of expected values (section 3.6),

$$\begin{aligned}
E(\bar{x}) &= \mu + \sum E(\varepsilon_i/n) \\
&= \mu + \sum 0 \\
&= \mu.
\end{aligned}$$

Thus $\bar{x}$ is an unbiased estimator of μ. Also

$$\begin{aligned}
\text{var}(\bar{x}) &= E(\bar{x}^2) - [E(\bar{x})]^2 \\
\bar{x}^2 &= (\mu + \sum \varepsilon_i/n)^2
\end{aligned}$$

$$= \mu^2 + 2\mu \sum \varepsilon_i/n + \sum \varepsilon_i^2/n^2 + \sum_{\substack{i \neq j}} \varepsilon_i \varepsilon_j/n^2 \ ;$$

$$E(\bar{x}^2) = \mu^2 + 2\mu \sum E(\varepsilon_i/n) + \sum E(\varepsilon_i^2/n^2)$$
$$+ \sum_{\substack{i \\ i \neq j}} \sum_{j} E(\varepsilon_i \varepsilon_j/n^2)$$
$$= \mu^2 + 2\mu \cdot 0 + \sum \sigma^2/n^2 + \sum\sum 0$$
$$= \mu^2 + \sigma^2/n.$$

Hence

$$\mathrm{var}(\bar{x}) = E(\bar{x}^2) - [E(\bar{x})]^2$$
$$= \mu^2 + \sigma^2/n - \mu^2$$
$$= \sigma^2/n.$$

Thus $\bar{x}$ is distributed with mean μ and variance σ^2/n. If the population sampled is normal, the distribution of $\bar{x}$ is normal. Even for a moderately skew or symmetric (but not normal) population the sampling distribution is approximately normal for large n—the larger n becomes, the closer the sampling distribution of the mean approximates to a normal distribution.

6.5 The Central Limit Theorem

The last sentence in the previous section—'even for a moderately skew... the sampling distribution of the mean approximates to a normal distribution'—follows from what is known as the *central limit theorem*. The proof of the theorem requires advanced mathematical methods and is not given here. However, a statement of it is:

The mean $\bar{x}$ of a sample of size n drawn from any population (continuous or discrete), with mean μ and finite variance σ^2, will have a distribution that approaches, as n becomes infinite, the normal distribution with mean μ and variance σ^2/n.

The restriction that the population sampled must have finite variance is of little practical significance since all but a few very special populations have finite variances.

At the beginning of the previous section, it was stated that the distribution of means from normal populations is of great importance to research workers in the agricultural and biological sciences. In many instances, however, the variables which have to be considered are not distributed exactly as normal variables. The significance

H

of this theorem is that the table of normal areas can be used to compute the probabilities of many statements concerning the arithmetic mean $\bar{x}$ even when the population sampled is not normal. Inferences about population means may be made even when the population is not normal. For skew distributions the approach to normality is quite rapid as the sample size is increased. Thus for many of the variables which biologists study, the means of relatively few observations may be taken to be normally distributed. It is for this reason that the normal distribution is used so much in biometrical applications.

6.6 Test of Significance for the Population Mean

The results contained in section 6.4 form the basis for many tests of significance.

Let $X_1, X_2, \ldots, X_n$ be the members of a random sample from a normal population with unknown mean μ and known variance σ^2, and let

$$\bar{x} = \sum_{i=1}^{n} X_i/n.$$

Suppose it is desired to test the hypothesis that the population mean is equal to μ_0 against the alternative that the mean was different from μ_0, i.e. to make the following test of significance:

Null hypothesis: $\qquad H_0: \mu = \mu_0,$
Alternative hypothesis: $\quad H_1: \mu \neq \mu_0.$
Level of significance $= \alpha$ (chosen arbitrarily by the experimenter).

If the null hypothesis is true, $\bar{x}$ is distributed normally with mean μ_0 and variance σ^2/n. Thus $z_0 = (\bar{x} - \mu_0)/(\sigma/\sqrt{n})$ is normally distributed with mean zero and unit standard deviation (i.e. z_0 is a standardized normal variate).

If, however, the null hypothesis is not true, and the population mean is μ_1 (where $\mu_1 = \mu_0 + \delta$) then $[\bar{x} - (\mu_0 + \delta)]/(\sigma/\sqrt{n})$ is a standardized normal variate and $(\bar{x} - \mu_0)/(\sigma/\sqrt{n})$ has a mean or expected value of $\delta\sqrt{n}/\sigma$.

The distribution of z_0 under the null hypothesis and two different alternative hypotheses are shown in Figure 6.1. Under the alternative hypothesis there is a tendency for more z_0 values different from zero to be obtained than when the null hypothesis is true. Values of z_0 greatly different from zero support the alternative hypotheses rather than the null hypothesis.

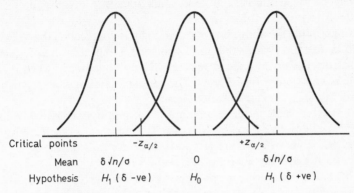

Critical points	$-z_{\alpha/2}$		$+z_{\alpha/2}$
Mean	$\delta\sqrt{n}/\sigma$	0	$\delta\sqrt{n}/\sigma$
Hypothesis	H_1 (δ -ve)	H_0	H_1 (δ +ve)

Fig. 6.1   Distribution of $z_0 = (\bar{x} - \mu_0)/(\sigma/\sqrt{n})$

The next step in the test of significance is to find the observed value z_{obs} of z_0. Then the test of significance may be completed in one of two ways.

(a) $P\{|z| \geqslant z_{obs}\}$ is obtained from the table of probabilities of the standardized normal variate. Suppose this probability is p. Then if $p < \alpha$, an unlikely value of z has been found, and as this was computed on the basis of the null hypothesis being correct, the null hypothesis is rejected.

(b) Alternatively, the significance level α and the table of the standardized normal are used to find the points $-z_{\frac{1}{2}\alpha}$ and $z_{\frac{1}{2}\alpha}$ which are such that $P\{|z| \geqslant z_{\frac{1}{2}\alpha}\} = \alpha$. These points are known as the *critical points* in the test of significance and the regions below $-z_{\frac{1}{2}\alpha}$ and above $z_{\frac{1}{2}\alpha}$ as the *critical regions*. If z_{obs} lies in either of these critical regions, it is concluded that an unlikely value of z has been found and as this was computed on the basis of the null hypothesis being correct, the null hypothesis is rejected.

If α is chosen to be 0.05, an observed value of z greater than 1.96 or less than -1.96 would be significant, while if α is chosen to be 0.01, z_{obs} would have to be such that it is either greater than 2.58 or less than -2.58 before the null hypothesis is rejected. The observed z is said to be *significant* or *highly significant* (depending on whether α is 5% or 1%) and it is conventional to indicate significance at the 5% level by a single asterisk and significance at the 1% level by a double asterisk (e.g. $z = 2.16^*$, $z = 3.14^{**}$).

In adopting the above line of reasoning it is appreciated that there will be instances when the null hypothesis is rejected when in fact it is true. The error made in rejecting the null hypothesis when it is true is known as a *type I error*. The probability of making a type I error is the level of significance chosen.

The error made in the acceptance of the null hypothesis when in fact an alternative hypothesis is true is known as a *type II error*. The probability of making a type II error depends on the level of significance and the alternative hypothesis.

In general, increasing the probability of a type I error decreases the probability of a type II error. As this is the case, an experimenter might wish to choose a level of significance other than the conventional 5% or 1%. For example, it might be thought to be desirable to minimize the sum of the probabilities of the two types of errors. This might be done if the null and alternative hypotheses were that the parameter under investigation had a particular value (e.g. $H_0 : \mu = 50$, $H_1 : \mu = 54$). However, problems of this kind are not elementary since in most agricultural and biological investigations, the alternative hypotheses are not of the simple type, $\mu = 54$, but are generally complex ones such as $\mu > 50$. When this is the case there is not a single distribution under the alternative hypothesis.

Up to this point, the test of significance which has been considered is what is known as a *two-tailed* test of significance. Since the alternative hypothesis was that $\mu \neq \mu_0$, observed values of z_0 which differed in either direction from the expected value of zero tended to support the alternative. If $z_{\text{obs}} < -z_{\frac{1}{2}\alpha}$ or $z_{\text{obs}} > z_{\frac{1}{2}\alpha}$, then this support of the alternative is so strong that the null hypothesis is rejected. Frequently, as indicated in the previous paragraph, the alternative hypothesis is that $\mu > \mu_0$. In this case, if $\bar{x}$ is found to be less than μ_0 there is no evidence to support the alternative hypothesis. However, if $\bar{x}$ is greater than μ_0 the evidence in support of the alternative hypothesis will require examination. The null hypothesis will be rejected if $z_{\text{obs}} > z_\alpha$ where z_α is such that $P\{z > z_\alpha\} = \alpha$. For $\alpha = 0.05$, z_α is 1.65. On the other hand, if the alternative hypothesis is

$\mu < \mu_0$, the test of significance is made if $\bar{x} < \mu_0$ and $-z_\alpha$ is chosen so that $P\{z < -z_\alpha\} = \alpha$. If $z_{obs} < -z_\alpha$, the null hypothesis is rejected. Such tests are called *one-tailed* tests of significance.

This introductory discussion on significance and hypothesis testing is presented here to indicate that there is nothing magical or awe-inspiring about the conventional 5% and 1% levels of significance, and there could be occasions when other levels of significance would be more appropriate.

Also instead of making a test of significance the experimenter may desire to infer from the sample data something about the value of the population parameter μ. This is an inductive process where the reasoning is from a part to the whole. Since $\bar{x}$ would vary from sample to sample, the experimenter would be loath to claim that a single observed value $\bar{x}$ is the value of μ. Using $\bar{x}$, it is better to compute an interval within which there is a large probability that μ would lie. This leads to the concept of confidence intervals—a topic which is introduced in the next section.

Example 6.1 The artificial nature of the following example will be immediately apparent. Rarely are biologists dealing with samples of 900, and still more rarely are they in the position of knowing the variance of the population from which they have taken their sample. The example is presented to illustrate the theory presented in the previous section. It should also be noted that use is made of the central limit theorem when it is stated that the distribution of $\bar{x}$ is approximately normal.

A sample of 900 members from a population whose distribution is symmetric and whose standard deviation is 2·61 cm is found to have a mean of 3·4 cm. Test the hypothesis that the population mean is 3·25 cm.

$$H_0 : \mu = 3\cdot25,$$
$$H_1 : \mu \neq 3\cdot25,$$

Level of significance: $\alpha = 0\cdot05$.

The s.e. of the distribution of $\bar{x}$ is

$$\sigma/\sqrt{n} = 2\cdot61/30 = 0\cdot087$$

and if the null hypothesis holds, the distribution of $\bar{x}$ has a mean

$\mu = 3\cdot25$. Further, since the sample is large, the distribution of $\bar{x}$ is approximately normal. Hence

$$\frac{(\bar{x} - \mu)}{(\sigma/\sqrt{n})}$$

is a standardized normal variate.
The observed value of z_0 is

$$z_{\text{obs}} = \frac{\bar{x} - \mu}{\sigma/\sqrt{n}} = \frac{3\cdot4 - 3\cdot25}{0\cdot087} = \frac{0\cdot15}{0\cdot087} = 1\cdot724.$$

Now since the alternative hypothesis is that $\mu \neq 3\cdot25$, either positive or negative values of z_0 which deviate greatly from zero will support the alternative hypothesis and throw doubt on the null hypothesis.

Again, since the level of significance has been chosen to be 5 %, the critical points $-z_{\frac{1}{2}\alpha}$ and $+z_{\frac{1}{2}\alpha}$ are $-1\cdot96$ and $1\cdot96$.

For this example, z_{obs} is neither less than $-1\cdot96$ nor greater than $1\cdot96$. Thus there is insufficient evidence to reject the hypothesis that this sample came from a population with mean of $3\cdot25$ cm.

6.7   Confidence Interval for Unknown Mean

Suppose that in the example above the null hypothesis had been that the population mean was $3\cdot23$ cm (or some value less than $3\cdot23$ cm), then the test of significance would have been significant and it would have been concluded that the sample did not come from such a population. Similarly, had the null hypothesis been that the population mean was $3\cdot57$ cm (or some value greater than $3\cdot57$ cm), a significant result would have been obtained. Only if a point in the interval between $3\cdot23$ cm and $3\cdot57$ cm is hypothesized for the population mean will a non-significant result be obtained. This interval is known as the 95% confidence interval for the population mean.

Consider now a random sample of size n drawn from a normal population with mean μ and standard deviation σ. Then the mean $\bar{x}$ is normally distributed with mean μ and standard deviation $\sigma/\sqrt{n}$.

If σ^2 is known but μ is not, there is a range of possible values which might be hypothesized for μ for which the observed mean $\bar{x}$

of the sample is not significant at a specified level of probability. For an observed $\bar{x}$ to be not significant at the α level of significance, $(\bar{x}-\mu)/(\sigma/\sqrt{n})$ must lie between the two critical points, $-z_{\frac{1}{2}\alpha}$ and $z_{\frac{1}{2}\alpha}$.

If the null hypothesis is true,

$$P\{-z_{\frac{1}{2}\alpha} \leqslant (\bar{x}-\mu)/(\sigma/\sqrt{n}) \leqslant z_{\frac{1}{2}\alpha}\} = 1-\alpha,$$

i.e.

$$P\{\bar{x} - z_{\frac{1}{2}\alpha}(\sigma/\sqrt{n}) \leqslant \mu \leqslant \bar{x} + z_{\frac{1}{2}\alpha}(\sigma/\sqrt{n})\} = 1-\alpha;$$

and if $\alpha = 0.05$ then

$$P\{\bar{x} - 1.96\sigma/\sqrt{n} \leqslant \mu \leqslant \bar{x} + 1.96\sigma/\sqrt{n}\} = 0.95. \tag{6.1}$$

The range of values $(\bar{x} - z_{\frac{1}{2}\alpha}\sigma/\sqrt{n})$ to $(\bar{x} + z_{\frac{1}{2}\alpha}\sigma/\sqrt{n})$ is the $(1-\alpha)$ *confidence interval* when σ is known. This is the range of values within which μ must be hypothesized to lie in order that the observed sample mean will not be significant at the α level of significance.

The statement contained in equation (6.1) that the probability is 0.95 that μ lies between $(\bar{x} - 1.96\sigma/\sqrt{n})$ and $(\bar{x} + 1.96\sigma/\sqrt{n})$ requires closer examination. It is not implied by this statement that μ is a variable. μ is a parameter and is fixed. The statement implies that if random samples were to be taken indefinitely from a normal population with unknown mean μ and known standard deviation σ and if for each of these samples the interval $(\bar{x} - 1.96\sigma/\sqrt{n})$ to $(\bar{x} + 1.96\sigma/\sqrt{n})$ were computed, then 95% of these intervals would contain the mean μ. 5% of such intervals would not cover the mean.

Definition. If it is possible to define two statistics g_1 and g_2 (functions of the sample values only) such that, θ being a parameter under estimate,

$$P\{g_1 \leqslant \theta \leqslant g_2\} = 1-\alpha$$

where α is some fixed probability, the interval between g_1 and g_2 is called the $(1-\alpha)$ *confidence interval*. The assertion that θ lies in this interval will be true, on the average, in a proportion $(1-\alpha)$ of the cases when the assertion is made.

Example 6.2 Show that, in Example 6.1, the 95% confidence interval for the mean is approximately 3.23 cm to 3.57 cm. (If the population sampled were normal, the probability associated with this interval would be exactly 0·95. Nevertheless with a sample size of 900, the probability would be very close to 0·95.)

The 95 % confidence interval is $\{\bar{x}-1.96\sigma/\sqrt{n}\}$ to $\{\bar{x}+1.96\sigma/\sqrt{n}\}$. Substituting for $\bar{x}$ and σ gives

$$\{3.4-(1.96\times2.61)/30\} \quad \text{to} \quad \{3.4+(1.96\times2.61)/30\},$$

i.e. $\{3.4-0.17\}$ to $\{3.4+0.17\}$, i.e. 3·23 to 3·57.

6.8 An Unbiased Estimator of σ^2

Many research workers in the biological sciences are interested not only in estimating the mean μ of the population from whence their sample has been drawn but also the variance σ^2 of the population. For example, an estimate of the variability of the concentration of vitamin B_{12} in the livers of lambs might be required; or, again, the variability of measurements from a particular chemical analysis.

Suppose a random sample of size n is drawn from a population with mean μ and variance σ^2. In section 6.4, $\bar{x}$ was shown to be an unbiased estimator of μ. It will now be shown that the sample variance s^2 is an unbiased estimator of σ^2 where

$$s^2 = \sum (X_i - \bar{x})^2/(n-1).$$

In finding $E[\sum (X_i - \bar{x})^2]$ and hence $E[\sum (X_i - \bar{x})^2/(n-1)]$, use is made of the relationship

$$\sum (X_i - \bar{x})^2 = \sum X_i^2 - (\sum X_i)^2/n.$$

The mathematical model is

$$X_i = \mu + \varepsilon_i, \quad (i = 1, 2, \ldots, n);$$
$$E(\varepsilon_i) = 0;$$
$$E(\varepsilon_i^2) = \sigma^2;$$
$$E(\varepsilon_i \varepsilon_j) = 0 \quad \text{if } i \neq j.$$

Now

$$\sum X_i = n\mu + \varepsilon_1 + \varepsilon_2 + \varepsilon_3 + \ldots + \varepsilon_n$$
$$= n\mu + \sum \varepsilon_i,$$
$$(\sum X_i)^2 = n^2\mu^2 + 2n\mu \sum \varepsilon_i + \sum \varepsilon_i^2 + \sum_{i \neq j} \varepsilon_i \varepsilon_j$$

where $\sum\limits_{i \neq j} \varepsilon_i \varepsilon_j$ is the sum of all product terms for which $i \neq j$, e.g.

terms of the type $\varepsilon_1 \varepsilon_2$ and $\varepsilon_3 \varepsilon_5$ but not ε_4^2, whence it can be shown that

$$E[(\textstyle\sum X_i)^2] = n^2\mu^2 + n\sigma^2;$$

hence

$$E[(\textstyle\sum X_i)^2/n] = n\mu^2 + \sigma^2. \tag{6.2}$$

Again

$$X_i^2 = \mu^2 + 2\mu\varepsilon_i + \varepsilon_i^2$$
$$\textstyle\sum X_i^2 = n\mu^2 + 2\mu \sum \varepsilon_i + \sum \varepsilon_i^2.$$

Therefore

$$\begin{aligned}
E(\textstyle\sum X_i^2) &= E(n\mu^2) + 2\mu \sum E(\varepsilon_i) + \sum E(\varepsilon_i^2) \\
&= n\mu^2 + 0 + n\sigma^2 \\
&= n\mu^2 + n\sigma^2.
\end{aligned} \tag{6.3}$$

Using equations (6.2) and (6.3),

$$\begin{aligned}
E[\textstyle\sum (X_i - \bar{x})^2] &= E[\textstyle\sum X_i^2 - (\sum X_i)^2/n] \\
&= E[\textstyle\sum X_i^2] - E[(\sum X_i)^2/n] \\
&= n\mu^2 + n\sigma^2 - n\mu^2 - \sigma^2 \\
&= (n-1)\sigma^2.
\end{aligned}$$

Therefore

$$E[\textstyle\sum (X_i - \bar{x})^2/(n-1)] = \sigma^2.$$

Thus s^2 is an unbiased estimator of σ^2.

It should be noted here that no reference is made to normality or otherwise of the population which is being sampled. The result that $E(s^2) = \sigma^2$ applies irrespective of the distribution type of the population.

From the above it might be assumed that s is an unbiased estimator of σ. That this is not so may be seen from the following.

It will be appreciated that in continued, repeated sampling the value of s^2 will vary from sample to sample. The statement that s^2 is an unbiased estimator of σ^2 is a short-hand way of saying that the average of all values of s^2 obtained in continued, repeated random sampling is σ^2. Since the values of s^2 are different

from sample to sample, the values of s will be different. Hence the sampling distribution of s has a variance.
Thus

$$\text{var}(s) > 0;$$

but

$$\text{var}(s) = E[s - E(s)]^2$$
$$= E(s^2) - [E(s)]^2,$$

hence

$$E(s^2) - [E(s)]^2 > 0$$
$$[E(s)]^2 < E(s^2)$$
$$< \sigma^2,$$

and on taking the square root of both sides of this relationship

$$E(s) < \sigma.$$

This proves that s is a biased estimator of σ. In actual fact this bias is very slight and s is frequently used as an estimator of σ.

Example 6.3   A random group of 7 six-week old chickens reared on a high protein diet weighed 10, 11, 15, 14, 13, 15, 13 ounces. Find an unbiased estimate of the variance of the population from which this group may be regarded as a random sample.

The group may be regarded as a random sample from the hypothetical infinite population of all possible six-week old chickens reared on the particular high protein diet.

An unbiased estimator of σ^2 is

$$s^2 = [\sum (X_i - \bar{x})^2]/(n-1).$$

Now $\sum X_i = 91$, and so

$$(\sum X_i)^2/n = 91^2/7$$
$$= 1183,$$
$$\sum X_i^2 = 10^2 + 11^2 + \ldots + 13^2$$
$$= 1205.$$

Hence

$$\sum (X_i - \bar{x})^2 = \sum X_i^2 - (\sum X_i)^2/n$$
$$= 1205 - 1183$$
$$= 22,$$

so that

$$s^2 = 22/6$$
$$= 3\cdot 67 \, (\text{oz})^2 .$$

6.9   Degrees of Freedom of s^2

The concept of degrees of freedom will be introduced briefly at this point. This concept presents difficulty in an elementary text of this type as it cannot be defined without a considerable knowledge of mathematics, e.g. the term degrees of freedom is most adequately defined in terms of the rank of a quadratic form.

Suppose a random sample of size n is drawn from a normal population (μ, σ^2). Then from Chapter 5, $\sum_{i=1}^{n} (X_i - \mu)^2/\sigma^2$ is distributed as a χ^2 with n degrees of freedom.

From section 6·4, $\bar{x}$ is normally distributed with mean μ and variance σ^2/n. Thus the difference of $\bar{x}$ from μ divided by the standard error of $\bar{x}$, i.e. $\sqrt{n}(\bar{x} - \mu)/\sigma$ is a standardized normal variate. Hence $n(\bar{x} - \mu)^2/\sigma^2$ is a χ^2 variate with one degree of freedom.

It is easy to show that

$$\frac{\sum (X_i - \mu)^2}{\sigma^2} = \frac{\sum (X_i - \bar{x})^2}{\sigma^2} + \frac{n(\bar{x} - \mu)^2}{\sigma^2}. \tag{6.4}$$

However it is not so readily shown, but nevertheless it is true, that $\sum (X_i - \bar{x})^2/\sigma^2$ is distributed as χ^2 with $(n-1)$ degrees of freedom.

Since $\sum (X_i - \bar{x})^2$ is distributed as $\sigma^2 \chi^2$ with $(n-1)$ degrees of freedom, $\sum (X_i - \bar{x})^2$ is spoken of as having $(n-1)$ *degrees of freedom*.

The relationship between $\sum (X_i - \bar{x})^2$ and the χ^2 distribution provides a neat proof that σ^2 is the expected value of s^2 when the population sampled is normal.

Since the expected value of a χ^2 variate with v degrees of freedom is v, and since $\sum (X_i - \bar{x})^2$ is distributed as $\sigma^2 \chi^2$ with $(n-1)$ degrees of freedom,

$$E[\sum (X_i - \bar{x})^2] = E(\sigma^2 \chi^2_{n-1})$$
$$= \sigma^2 E(\chi^2_{n-1})$$
$$= \sigma^2 (n-1).$$

Hence

$$E[\sum (X_i - \bar{x})^2/(n-1)] = \sigma^2$$
$$E(s^2) = \sigma^2.$$

The proof outlined in section 6.8 is, of course, more general than this since it applies whether the population is normal or not.

Exercises

6.1   A random sample of 400 members from a population whose variance is 16 is found to have a mean of 30. Find the 90% confidence interval for the mean and hence test the hypothesis (at the 10% level of significance) that the population mean is 30·5.

6.2   The figures below are for protein tests on nine samples of a variety of wheat grown in a particular district. On the assumption that protein determinations are normally distributed with standard deviation 0·54, find the 95% confidence interval for the mean. If the variance were not known, what would be an unbiased estimate of it from the data of the sample?

    12·9, 13·4, 12·4, 12·8, 13·0, 12·7, 12·4, 13·5, 13·9.

6.3   The data for a random sample of five from an infinite, continuous population were 13·1, 12·7, 13·3, 12·9 and 13·0. Find an unbiased estimate of the population variance.

6·4(a)   What is the difference between σ and s?

   (b)   The values 14, 16, 11, 12, 10, 14, 10, 12, 15 are drawn at random from a normal population of known variance equal to 4 and of unknown mean. Estimate the mean of the population and find the 95% confidence interval.

6.5   From a population of grasshoppers a sample of nine is chosen at random. If the weights of the grasshoppers are normally distributed with unknown mean and a known standard deviation of 0·3 g, what is the probability that the sample mean ($\bar{x}$) exceeds the population mean by more than 0·29 g?

6.10   Confidence Interval and a Test of Significance for the Variance of a Normal Population

The fact that $\sum (X_i - \bar{x})^2/\sigma^2$, i.e. $(n-1)s^2/\sigma^2$, is distributed as χ^2

with $(n-1)$ degrees of freedom is used to obtain confidence intervals for the variance of a normal population.

If $\chi^2_{(1-\frac{1}{2}\alpha),(n-1)}$ and $\chi^2_{\frac{1}{2}\alpha,(n-1)}$ are such that

$$P\{\chi^2_{(n-1)} \leqslant \chi^2_{(1-\frac{1}{2}\alpha),(n-1)}\} = \tfrac{1}{2}\alpha,$$
$$P\{\chi^2_{(n-1)} \geqslant \chi^2_{\frac{1}{2}\alpha,(n-1)}\} = \tfrac{1}{2}\alpha,$$

then

$$P\{\chi^2_{(1-\frac{1}{2}\alpha),(n-1)} \leqslant \chi^2_{(n-1)} \leqslant \chi^2_{\frac{1}{2}\alpha,(n-1)}\} = \alpha,$$
$$P\{\chi^2_{(1-\frac{1}{2}\alpha),(n-1)} \leqslant (n-1)s^2/\sigma^2 \leqslant \chi^2_{\frac{1}{2}\alpha,(n-1)}\} = \alpha,$$
$$P\{(n-1)s^2/\chi^2_{\frac{1}{2}\alpha,(n-1)} \leqslant \sigma^2 \leqslant (n-1)s^2/\chi^2_{(1-\frac{1}{2}\alpha),(n-1)}\} = 1-\alpha.$$

This final line may be obtained by algebraic manipulation of the previous line. Thus the $(1-\alpha)$ confidence interval is

$$\{(n-1)s^2/\chi^2_{\frac{1}{2}\alpha,(n-1)}\} \quad \text{to} \quad \{(n-1)s^2/\chi^2_{(1-\frac{1}{2}\alpha),(n-1)}\}.$$

In particular the 95 % confidence interval is

$$\{(n-1)s^2/\chi^2_{0\cdot025,(n-1)}\} \quad \text{to} \quad \{(n-1)s^2/\chi^2_{0\cdot975,(n-1)}\}.$$

s^2 is calculated from the sample data and $\chi^2_{0\cdot025,(n-1)}$ and $\chi^2_{0\cdot975,(n-1)}$ read from the χ^2 table. Since $\chi^2_{0\cdot025,(n-1)}$ is the value of χ^2 which is exceeded with probability $0\cdot025$ it is much larger than $\chi^2_{0\cdot975,(n-1)}$, this later being the value of χ^2 exceeded with probability $0\cdot975$. It will be realized that the probability of obtaining a χ^2 between $\chi^2_{0\cdot025}$ and $\chi^2_{0\cdot975}$ is $0\cdot95$.

The $(1-\alpha)$ confidence interval may be used to make a two-tailed test of significance at the α level of significance. If the confidence interval contains the hypothesized value for the variance then the null hypothesis is retained; otherwise, it is rejected. Alternatively, if the null hypothesis is that $\sigma^2 = \sigma_0^2$, then $(n-1)s^2/\sigma_0^2$ can be calculated and the observed χ^2 compared with those tabulated in the χ^2 table. Just as in the case of the mean, one- and two-tailed tests of significance can be made, the alternative hypothesis determining the critical points and critical regions.

Example 6.4 In Example 6.3 an unbiased estimate s^2 of the population variance σ^2 was obtained from a sample of 7 six-week old chickens. Assuming that the weights of six-week old chickens are normally distributed, the 95% confidence interval for σ^2 is obtained as follows.

$$\chi^2_{0.025,6} = 14.45,$$
$$\chi^2_{0.975,6} = 1.24.$$

Since $\qquad s^2 = 3.67$ and $(n-1)s^2 = 22,$

the 95% confidence interval is $(22/14.45)$ to $(22/1.24)$, i.e. 1.522 to 17.741.

This interval may be used to test the hypothesis

$$H_0 : \sigma^2 = 10,$$
$$H_1 : \sigma^2 \neq 10,$$
$$\alpha = 0.05.$$

It is noted that the confidence interval (1.522 to 17.741) contains 10 and thus the null hypothesis is not rejected.

Alternatively

$$(n-1)s^2/\sigma^2 = 22/10 = 2.2.$$

This observed χ^2 of 2.2 is neither less than 1.24 nor greater than 14.45 and so the null hypothesis is not rejected. 1.24 and 14.45 are the two critical points for this two-tailed test of significance. They are obtained by entering Table II with six degrees of freedom.

6.11 The Use of the χ^2 Distribution for Confidence Interval Statements and Tests of Significance concerning the Mean

The result obtained in section 6.9 that $[(\bar{x}-\mu)/(\sigma/\sqrt{n})]^2$ is a χ^2 variate with one degree of freedom may be used to calculate confidence intervals or to make tests of significance concerning the mean. Instead of using the z distribution to make the test of significance, $[(\bar{x}-\mu_0)/(\sigma/\sqrt{n})]^2$ can be calculated and the observed value compared with the χ^2 variate with one degree of freedom. Similarly, confidence interval statements might be made using this relationship.

Examination of the z and χ^2 tables shows that the entries in the normal table are the square root of those in the χ^2 for one degree of freedom, e.g. $\chi^2_{0.05,1} = 3.84$ which is the square of 1.96.

6.12 The Additive Property of χ^2

Examination of equation (6.4) shows that the left-hand side of the equation is a χ^2 variate with n degrees of freedom, while the two terms on the right-hand side are χ^2 variates with $(n-1)$ and one degree of freedom respectively. Thus

$$\chi_n^2 = \chi_{n-1}^2 + \chi_1^2$$

This is an illustration of what is known as the additive property of χ^2. This property, which is almost obvious from the definition of χ^2, might be stated formally as follows:

If $\chi_1^2, \chi_2^2, \chi_3^2, \ldots, \chi_k^2$ are k independently distributed χ^2 variates with $v_1, v_2, v_3, \ldots, v_k$ degrees of freedom respectively, their sum

$$\chi_1^2 + \chi_2^2 + \chi_3^2 + \ldots + \chi_k^2$$

is a χ^2 variate with degrees of freedom

$$v_1 + v_2 + v_3 + \ldots + v_k.$$

One effect of this is that while the results of single experiments may be of doubtful significance, the χ^2 test for the pooled data gives due weight to the cumulative evidence of the separate experiments.

Example 6.5 Three experiments designed to test a hypothesis concerning the variance of a population yielded

$$\chi_1^2 = 9.00 \quad \text{with } v_1 = 5,$$
$$\chi_2^2 = 16.12 \quad \text{with } v_2 = 10,$$
$$\chi_3^2 = 22.31 \quad \text{with } v_3 = 15.$$

None of these values on its own is significant at the 5% level. However, the sum $\chi^2 = 47.43$ for $v = 30$ is significant at this level.

Thus while each experiment on its own gave a non-significant result, the pooled data gave a value of χ^2 which, should the hypothesis be correct, would be expected due to chance with probability 0.02.

Example 6.6 Fisher (1950) has shown that $-2 \ln p^*$, (where p is the probability of obtaining a value of a test criterion as extreme as,

---

* $\ln \equiv$ logarithm to base e.

or more extreme than, that obtained in a particular test), is distributed as χ^2 with two degrees of freedom. By the additive property of χ^2 such values may be added. Thus if $\chi_1^2, \chi_2^2, \chi_3^2$ are the values of χ^2 for $v = 2$ corresponding to p_1, p_2, p_3, the pooled value of χ^2 is

$$-2(\ln p_1 + \ln p_2 + \ln p_3)$$

with

$$v = 2+2+2 = 6 \text{ d.f.}$$

From the χ^2 table the combined probability may then be obtained. If $p_1 = 0.350$, $p_2 = 0.150$, $p_3 = 0.250$, then

$$\begin{aligned}
\chi^2 &= -2(\ln 0.350 + \ln 0.150 + \ln 0.250) \\
&= -2(-1.04981 - 1.89712 - 1.38629) \\
&= -2(-5.33322) \\
&= 10.66644.
\end{aligned}$$

The χ^2 table shows $P(\chi^2_{6 \text{ d.f.}} \geq 10.655) = 0.10$. Thus the combined probability is approximately 10%.

Exercises

6.6   A certain hypothesis was tested by three similar experiments. These gave $\chi^2 = 11.9$ with $v = 6$, $\chi^2 = 14.2$ with $v = 8$, and $\chi^2 = 18.3$ with $v = 11$. Show that the three experiments together provide more justification for rejecting the hypothesis than any one experiment alone.

6.7   A certain hypothesis was tested by two similar experiments which gave $\chi^2 = 14.7$ with $v = 9$ and $\chi^2 = 14.9$ with $v = 11$. Show that the two experiments combined give less reason for confidence in the hypothesis than either experiment alone.

6.13   Sampling of Qualitative Variables

In the sampling of attributes, as distinct from the sampling of continuous variables such as height, the possession or non-possession of some specified attribute or characteristic is of primary concern. For example, in sampling a population of pea seeds the experimenter may be concerned only with whether a seed is wrinkled or smooth.

In problems of this type the selection of a simple sample of n members is identical with that of a series of n independent trials, with constant probability π of success. Thus the results of Chapter 4 on the binomial distribution are applicable. The probabilities of $0, 1, 2, \ldots, n$ individuals possessing the particular characteristic in a simple sample of n individuals are the terms in the binomial distribution with parameters n and π. This is the distribution which would be obtained by continued repeated sampling, and the binomial probability distribution thus determined is called the *sampling distribution* or *derived distribution* of the observed number. The mean of this distribution or expected value is $n\pi$ and the standard deviation is $\sqrt{(n\pi(1-\pi))}$. Hence the standard error of the observed number is $\sqrt{(n\pi(1-\pi))}$.

The proportion of individuals in the sample possessing the attribute is obtained by dividing the observed number by the sample size n. The expected value of the observed proportion p is thus $n\pi/n = \pi$. Hence, p is an unbiased estimator of π.

Since the variance of the observed number is $n\pi(1-\pi)$, the variance of the proportion (number $\div n$) is $\pi(1-\pi)/n$ and the standard error (s.e.) of the proportion is $\sqrt{(\pi(1-\pi)/n)}$.

6.14   Large Sample Test of Significance and Confidence Interval for Population Proportion

In the preceding section it has been shown that the sampling distributions of the number and proportion of individuals with a particular characteristic in a simple sample of size n are binomial distributions. Also it has been stated in Chapter 5 that for large values of n the binomial distribution approximates to a normal distribution.

Thus from a knowledge of the normal distribution, a test can be made of the hypothesis that a given large simple sample of n members was obtained from a population in which the relative frequency of the occurrence of the attribute under investigation is π_0.

Let X be the observed number of individuals with the specified characteristic in the simple sample of size n.

Null hypothesis: $\quad H_0 : \pi = \pi_0$,
Alternative hypothesis: $\quad H_1 : \pi \neq \pi_0$.
Level of significance: $\quad \alpha$

(the latter chosen arbitrarily by the experimenter).

I

Then if the null hypothesis is true,

$$\text{Mean of } X \text{ or } E(X) = n\pi_0,$$
$$\text{Variance } (X) = n\pi_0(1-\pi_0).$$

Since n is large, X is approximately normally distributed. Thus, $X - n\pi_0$ is approximately normally distributed with mean zero and variance $n\pi_0(1-\pi_0)$.

$$z_0 = \frac{X - n\pi_0}{\sqrt{(n\pi_0(1-\pi_0))}}$$

is approximately normally distributed with mean zero and unit standard deviation.

To complete the test of significance the observed value of z_0 (say z_{obs}) is found and $P\{z > |z_{obs}|\} = p/2$ is found from Table I.

If this probability is small and $p < \alpha$, the null hypothesis is rejected. It is concluded that π is not equal to π_0. If, however, $p > \alpha$, the null hypothesis stands until further experimental evidence is available.

Alternatively, the critical points $-z_{\frac{1}{2}\alpha}$ and $z_{\frac{1}{2}\alpha}$ may be read from the z table and z_{obs} compared with these. If $z_{obs} < -z_{\frac{1}{2}\alpha}$ or $z_{obs} > z_{\frac{1}{2}\alpha}$ the null hypothesis will be rejected.

The preceding test is a two-tailed test of significance. If required a one-tailed test might be made; here, of course, there would be a single critical region.

It will be appreciated that the above argument holds if in place of the number (X) and its standard error $\sqrt{(n\pi(1-\pi))}$, the proportion $(p = X/n)$ and its standard error $\sqrt{(\pi(1-\pi)/n)}$ are used, i.e.

$$\frac{(X/n) - \pi_0}{\sqrt{(\pi_0(1-\pi_0)/n)}}$$

is a z variate.

If an approximate confidence interval for the parameter π is required, this may be obtained from a knowledge of the normal distribution and by using the sample estimate $\hat{\pi}$ in place of π in the formula for the standard error. Thus since π is unknown the standard error of the sampling distribution is taken to be $\sqrt{(\hat{\pi}(1-\hat{\pi})/n)}$.

However, since the sample size is large, the difference between $\sqrt{(\pi(1-\pi)/n)}$ and $\sqrt{(\hat{\pi}(1-\hat{\pi})/n)}$ is not very great. For example, for a sample of size 100, if $\pi = 0.5$, then

$$\sqrt{(\pi(1-\pi)/n)} = 0.05.$$

If now $\hat{\pi} = 0.4$, then

$$\sqrt{(\hat{\pi}(1-\hat{\pi})/n)} = 0.049.$$

Even if $\hat{\pi}$ were 0.3,

$$\sqrt{(\hat{\pi}(1-\hat{\pi})/n)} = 0.046.$$

Thus the $(1-\alpha)$ approximate confidence interval for π is

$$\{\hat{\pi} - z_{\frac{1}{2}\alpha}\sqrt{(\hat{\pi}(1-\hat{\pi})/n)}\} \text{ to } \{\hat{\pi} + z_{\frac{1}{2}\alpha}\sqrt{(\hat{\pi}(1-\hat{\pi})/n)}\}.$$

The discussion in this section has centred around significance tests and confidence intervals when the sample size is large. The methods developed are also appropriate when n and π are such that $n\pi > 10$ and $n\pi(1-\pi) > 10$. However if either $n\pi$ or $n\pi(1-\pi)$ is less than 10, the terms in the binomial distribution do not agree well with the corresponding areas under the normal curve. Thus when either $n\pi < 10$ or $n\pi(1-\pi) < 10$ the actual terms in the binomial distribution should be used in any test of significance or confidence interval statement. Tables for such tests and intervals have been prepared, e.g. Owen (1962), while the papers by Clopper and Pearson (1934) and Crow (1956) would serve as an introduction to the literature on this problem.

6.15 Test of Significance of Difference of Two Proportions

Suppose an experimenter finds that in a simple sample of 600 wheat plants from a certain large district, 200 are affected by a particular disease. In one of 800 from another large district, again 200 are affected. Suppose the experimenter desires to determine whether the data indicate that the districts are significantly different with respect to the prevalence of the disease.

In order to solve a problem such as this, consider two populations, P_1 and P_2, which are sampled for the prevalence of a certain attribute by taking from them large simple samples of n_1 and n_2 members respectively and let p_1 and p_2 be the observed proportions in the two samples. Thus p_1 and p_2 are estimates of population parameters, π_1 and π_2.

On the hypothesis that $\pi_1 = \pi_2 (= \pi)$, the two samples may be combined to estimate the common value (π) of the relative frequency of the occurrence of the attribute in the populations. This estimate is written as $\hat{\pi}$ and is calculated as follows.

$$\hat{\pi} = p = \frac{n_1 p_1 + n_2 p_2}{n_1 + n_2}$$

$$= \frac{\text{Number with attribute in the two samples}}{\text{Total number in the two samples}}.$$

Estimates of variances of the proportions in the two samples of n_1 and n_2 members are $p(1-p)/n_1$ and $p(1-p)/n_2$ respectively. Since the samples are independent, an estimate of the variance of the difference of these proportions is given by

$$\hat{\sigma}_D^2 = \frac{p(1-p)}{n_1} + \frac{p(1-p)}{n_2}$$

$$= p(1-p) \left[\frac{1}{n_1} + \frac{1}{n_2} \right].$$

Again on the hypothesis that the populations have the same proportion, that is $\pi_1 = \pi_2 = \pi$,

$$E(p_1) = \pi, \qquad E(p_2) = \pi.$$

Hence

$$E(p_1 - p_2) = E(p_1) - E(p_2)$$

$$= \pi - \pi = 0.$$

The sampling distributions of p_1 and p_2 are approximately normal when n_1 and n_2 are large. The same is true of their difference.

Thus the distribution of $p_1 - p_2$ is approximately normal with mean zero and standard error, σ_D. Just as in the previous section when the confidence interval had to be calculated, an estimate of the standard error of the sampling distribution has to be used.

Thus $(p_1 - p_2)/\hat{\sigma}_D$ is considered to be distributed as z. If the null hypothesis is true the probability of the observed difference, $p_1 - p_2$, being less than $-2 \cdot 58 \hat{\sigma}_D$ or greater than $2 \cdot 58 \hat{\sigma}_D$ is very nearly 1%. The probability that it will be less than $-1 \cdot 96 \hat{\sigma}_D$ or greater than

$1.96\hat{\sigma}_D$ is approximately 5%. These will be the values used if tests are made at the 1% or 5% level of significance respectively.

For those who wish to calculate a confidence interval instead of making a test of significance, the $(1-\alpha)$ approximate confidence interval for $\pi_D(=\pi_1-\pi_2)$ is

$$\{(p_1-p_2)-z_{\frac{1}{2}\alpha}\hat{\sigma}_D\} \quad \text{to} \quad \{(p_1-p_2)+z_{\frac{1}{2}\alpha}\hat{\sigma}_D\}.$$

Having noted earlier the relationship between the z and χ^2 distributions, it is interesting to point out before leaving these problems that the two tests of significance in this and the previous section can be made by a χ^2 test of goodness of fit. This use of χ^2 is introduced in Chapter 11.

Example 6.7 A random sample of 500 pineapples was taken from a very large consignment and 75 were found to be bad. Show that an estimate of the standard error of the proportion of bad pineapples in a sample of this size is 0.016. Test the hypothesis, at the 1% level, that π equals 0.10.

From the sample data, an estimate of π is

$$p = 75/500 = 0.15.$$

Estimated s.e. of proportion of bad pineapples is

$$\sqrt{(p(1-p)/n)} = \sqrt{(0.15 \times 0.85/500)}$$
$$= 0.16.$$

Now

$$H_0 : \pi = 0.10,$$
$$H_1 : \pi \neq 0.10,$$
$$\alpha = 0.01.$$

The observed proportion $p = 0.15$ and

$$E(p) = \pi = 0.10.$$

Hence

$$z_0 = \frac{p-0.10}{\sqrt{((0.10)(0.90)/500)}}$$

is a standardized normal variate.

$$z_{obs} = \frac{0 \cdot 05}{0 \cdot 013}$$

$$= 3 \cdot 85.$$

With $\alpha = 0 \cdot 01$, the critical points of z are $-2 \cdot 58$ and $2 \cdot 58$.

Since $3 \cdot 85$ is greater than $2 \cdot 58$, the null hypothesis is rejected. It is concluded that the proportion of bad pineapples in the consignment is different from 10%.

Example 6.8 In a simple sample of 600 wheat plants from a certain large district, 200 are found to be affected by a particular disease. In one of 800 from another large district, 200 are found to be affected. Do the data indicate that the districts are significantly different with respect to the prevalence of the disease? Test the hypothesis that $\pi_1 = \pi_2 (= \pi)$ and find the 99% confidence interval for the difference.

Here

$$p_1 = 200/600 = 0 \cdot 333,$$
$$p_2 = 200/800 = 0 \cdot 25,$$
$$p_1 - p_2 = 0 \cdot 083.$$

On the hypothesis that the districts are alike with respect to the prevalence of disease, that $\pi_1 = \pi_2 (= \pi)$, an estimate of the parameter π is

$$\hat{\pi} = p = \frac{\text{Number of diseased plants in two samples}}{\text{Total number of plants in two samples}}$$

$$= \frac{200 + 200}{600 + 800}$$

$$= 4/14.$$

An estimate of the variance of the difference of the proportion for the two samples is

$$\hat{\sigma}_D^2 = p(1-p)\left[\frac{1}{n_1} + \frac{1}{n_2}\right]$$

$$= \frac{4}{14} \times \frac{10}{14}\left[\frac{1}{600} + \frac{1}{800}\right]$$

$$= \frac{4 \times 10 \times 14}{14 \times 14 \times 4800}$$

$$= 0.000595.$$

Hence

s.e. of difference $= 0.024$

$$z_{obs} = (p_1 - p_2)/(\text{s.e. of difference})$$

$$= 0.083/0.024 = 3.46.$$

This observed value of z is greater than the critical points either at the 5% or 1% levels of significance. There is less than one chance in a hundred if the districts were affected equally by this disease, that two such samples would have as large a difference in their observed proportions.

The observed difference is highly significant and the null hypothesis is rejected at the 1% level.

An approximate 99% confidence interval is

$$\{(p_1 - p_2) - 2.58\hat{\sigma}_D\} \quad \text{to} \quad \{(p_1 - p_2) + 2.58\hat{\sigma}_D\};$$
$$\{(p_1 - p_2) - 2.58\hat{\sigma}_D\} = 0.083 - 2.58 \times 0.024 = 0.022,$$
$$\{(p_1 - p_2) + 2.58\hat{\sigma}_D\} = 0.083 + 2.58 \times 0.024 = 0.144.$$

This interval $\{0.022$ to $0.144\}$ does not contain zero which is the value hypothesized for the difference in the two population proportions. Since a confidence interval is the range of values which might be hypothesized for a parameter and still get a non-significant test of significance, the fact that this interval does not contain zero implies the rejection at the 1% level of the null hypothesis that π_1 and π_2 are equal.

Exercises

6.8 Two insecticides are to be tested for their efficiency to kill insects. The efficiency is to be assessed by the proportion of insects which are killed in a given time. A random sample of 500 insects is sprayed with the first insecticide and 400 die in the specified time. Another random sample of 500 insects is sprayed with the second insecticide and 320 die in the same period of time. Test the hypothesis at the 1% level of significance that the insecticides are equally efficient.

6.9 Repeat Exercise 6.8 using the data of Table 1.1. This was one of the first problems posed in the first chapter.

6.10 The germination rate of lettuce seeds after two different methods of seed-bed preparation are compared by sowing random samples of 400 seeds after the first method and 300 seeds after the second method. The percentage of seeds germinating with the first method is 80% and under the second method 85%. Find the 90% confidence interval for the observed difference and hence test, at the 10% level of significance, the hypothesis that the proportions of germinating seeds are the same after both methods.

REFERENCES

CLOPPER, C. J. and PEARSON, E. S. (1934). 'The Use of Confidence or Fiducial Limits Illustrated in the Case of the Binomial', *Biometrika*, **26**, 404–413.

CROW, E. L. (1956). 'Confidence Intervals for a Proportion', *Biometrika*, **43**, 423–435.

FISHER, R. A. (1950). *Statistical Methods for Research Workers* (11th edition). Oliver & Boyd, Edinburgh; Hafner, New York.

OWEN. D. B. (1962). *Handbook of Statistical Tables*. Addison–Wesley, Reading, Mass.

COLLATERAL READING

EHRENFELD, S. and LITTAUER, S. B. (1964). *Introduction to Statistical Method*. McGraw-Hill, New York. Chapter 7.

FINNEY, D. J. (1964). *An Introduction to Statistical Science in Agriculture* (2nd edition). Oliver & Boyd, Edinburgh; Hafner, New York. Chapter 8.

FRASER, D. A. S. (1958). *Statistics—An Introduction*. John Wiley, New York. Chapters 9 and 10.

GOLDSTEIN, A. (1964). *Biostatistics—An Introductory Text*. Macmillan, New York. Chapter 2.

LI, J. C. R. (1964). *Statistical Inference*, Vol. I. Edwards Brothers, Ann Arbor, Michigan. Chapters 4, 5, 6, 11.

LI, C. C. (1964). *Introduction to Experimental Statistics*. McGraw-Hill, New York. Chapters 4 and 6.

STEEL, R. G. D. and TORRIE, J. H. (1960). *Principles and Procedures of Statistics*. McGraw-Hill, New York. Chapters 3 & 4.

WEATHERBURN, C. E. (1947). *A First Course in Mathematical Statistics*. Cambridge University Press, Cambridge. Chapter 6.

CHAPTER 7

The t Distribution

7.1 Introduction

In the previous chapter the distribution of $z = (\bar{x} - \mu)/(\sigma/\sqrt{n})$ was used in tests of significance and interval estimation concerning the population mean. In the type of problem which was considered there, the population variance was assumed known and it did not have to be estimated. However, in biological research the population variance is usually not known and an unbiased estimate s^2, obtained from the sample data, has to be used in place of σ^2. Then the distribution

$$t = \frac{\bar{x} - \mu}{s/\sqrt{n}}$$

has to be used in significance testing and interval estimation. The t distribution was first found by W. S. Gosset (1876–1937) who wrote under the *nom de plume* of 'Student'. For this reason the distribution has come to be known as Student's t distribution.

The t distribution is not normal. The parameter of the distribution is the number of degrees of freedom (v) appropriate to the estimate of the population variance. Thus in solving a particular problem, the degrees of freedom for s^2 determine the actual t distribution which has to be used.

The probability curves for t are symmetrical. As may be seen from Figure 7.1, a t curve is somewhat flatter than that for a standardized normal distribution, lying under it at the centre and above it at the tails. Also, as the degrees of freedom increase the t distribution approaches the normal.

Table III in the appendix gives percentage points $t_{\frac{1}{2}\alpha}$ of the t distribution. In contrast with Table I, the t table gives t values rather than probabilities in the body of the table. The t table is entered for degrees of freedom appropriate to the estimate of the population variance. At the top of Table III are probabilities (α) and in the body the corresponding values of $t_{\frac{1}{2}\alpha}$ which are such that $P(t \geqslant t_{\frac{1}{2}\alpha}) = \frac{1}{2}\alpha$. For example, on entering the table at the line with degrees of freedom $v = 8$, and in the column headed $\alpha = 0.05$ is found $P(t_{8} \geqslant 2.306) = 0.05/2 = 0.025$. Because of the symmetry of the t

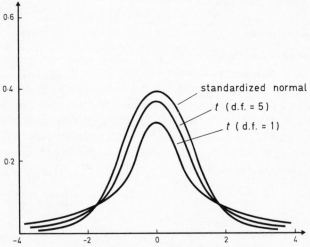

Fig. 7.1   Two t distributions and a standardized normal curve

distribution $P\{(t_8 \leqslant -2.306) \text{ or } (t_8 \geqslant 2.306)\} = 0.05$. Thus the probabilities stated at the top of the table correspond to the areas under both tails of the probability curve. It is important to remember this characteristic of the t table because if the area under one tail only is required, the probabilities as tabulated in Table III should be halved.

The approach to normality which is indicated by Figure 7.1 can be seen from an examination of Table III since the entries in the last line, $v = \infty$, are those for a standardized normal distribution and the entries in any column are obviously approaching the entry in the last line.

Exercises

7.1   Find t_0 such that $P(t \geqslant t_0) = 0.025$ for 10 d.f., $P(t \leqslant t_0) = 0.01$

for   12d.f., $P\{(t \leqslant -t_0)$ or $(t \geqslant t_0)\} = 0.10$ for   20d.f.,
$P(-t_0 \leqslant t \leqslant t_0) = 0.70$ for 30 d.f.

7.2   Find $P(-1.1 \leqslant t \leqslant 3.25)$ for 9 d.f., $P(0 \leqslant t \leqslant 2.16)$ for 13 d.f.

7.2   Test for an Assumed Population Mean

One use of the statistic t is to test the null hypothesis that the mean
of a normal population (of unknown variance) is, say, μ_0. A random
sample of size n is drawn from the population. The statistics $\bar{x}$ and s
are calculated from the sample and then

$$t = \frac{\bar{x} - \mu_0}{s/\sqrt{n}}. \tag{7.1}$$

The level of significance determines the critical points (or critical
point if a one-tailed test is being made) of the t distribution. These
are read from Table III after entering the table with the degrees of
freedom of s^2. If the observed t lies in the critical regions the null
hypothesis is rejected; if not, the null hypothesis stands. Alternatively,
Table III may be used to obtain the probability of the observed t
being exceeded numerically in random sampling from a normal
population with mean μ_0. If this probability is less than the chosen
level of significance the null hypothesis is rejected; otherwise, it
stands.

In the following example, the four steps in the test of significance
(as listed in section 1.3) are clearly indicated. While it is not usual to
indicate these steps as clearly as is done in the example, it should be
remembered that all tests of significance have them.

Example 7.1   The figures below are for protein tests on nine
samples of a variety of wheat grown in a particular district. Given
that protein determinations are approximately normally distributed,
show that the data are consistent with the assumption of a mean
protein of 12·66 for the district.

12·9, 13·4, 12·4, 12·8, 13·0, 12·7, 12·4, 13·5, 13·9.

Step 1.   Null hypothesis: $H_0 : \mu = 12.66$,
               Alternative hypothesis: $H_1 : \mu \neq 12.66$.

Step 2.   Level of significance $= 0.05$.

Step 3.   From the sample,

$$n = 9, \sum X = 117.0, \ \bar{x} = 13.0 \text{ and}$$
$$\sum X^2 = 1523.08,$$
$$\text{C.T.} = (\sum X)^2/9 = 1521.00 \,;$$
$$\sum (X - \bar{x})^2 = \sum X^2 - (\sum X)^2/9 = 2.08 \,;$$
$$s^2 = \tfrac{1}{8} \sum (X - \bar{x})^2$$
$$= 0.26,$$
$$s = 0.51.$$

Hence

$$t = \frac{\bar{x} - \mu}{s/\sqrt{n}} = \frac{13.0 - 12.66}{0.51/\sqrt{9}}$$
$$= \frac{13.0 - 12.66}{0.17} = \frac{0.34}{0.17}$$
$$= 2.00.$$

If the null hypothesis is correct, Table III shows (on entering it with $v = 8$ d.f.) that a value of $t \leqslant -2.00$ or $\geqslant 2.00$ would be obtained with a probability of about 8%.

Step 4.   Since $p(\doteqdot 0.08) >$ level of significance $\alpha(= 0.05)$, the null hypothesis is not rejected, and it is concluded that the data are not inconsistent with a mean protein of 12.66 for the district.

This final step might have been made by using the Table III to obtain the 5% critical points. For $v = 8$ and a two-tailed test of significance, these are -2.306 and 2.306. The observed t of 2.00 is neither < -2.306 nor > 2.306 and thus the null hypothesis is not rejected.

It might be noted that if the null hypothesis had been that $\mu = 12.608$ or $\mu = 13.392$, a value of $t = -2.306$ or $t = 2.306$ would have been obtained, the t value would have been significant and the null hypothesis rejected. Similarly, for all values of $\mu \leqslant 12.608$ or $\mu \geqslant 13.392$, a significant value of t would be obtained.

7.3   Confidence Intervals for the Population Mean

This problem is similar to that in section 6.6; the difference is that here the population variance is not assumed known.

As there, the $(1-\alpha)$ *confidence interval* for the population mean μ is defined as that range of values for μ which, with the data of the sample, will furnish a non-significant t at the α level of significance.

If $-t_{\frac{1}{2}\alpha}$ and $t_{\frac{1}{2}\alpha}$ are the critical points, then for non-significance $(\bar{x}-\mu)/(s/\sqrt{n})$ must lie between $-t_{\frac{1}{2}\alpha}$ and $t_{\frac{1}{2}\alpha}$. Thus for a non-significant t test, μ must be hypothesized to be such that

$$\bar{x}-t_{\frac{1}{2}\alpha}(s/\sqrt{n}) \leqslant \mu \leqslant \bar{x}+t_{\frac{1}{2}\alpha}(s/\sqrt{n}).$$

This range of values is the $(1-\alpha)$ confidence interval. The end points of the interval are the $(1-\alpha)$ *confidence limits*. 99% and 95% confidence intervals and limits are calculated when the 1% and 5% critical points are used.

In contrast with section 6.6 the points on the t distribution are used here, whereas when the variance is known the z distribution is used.

Also, as stated in the previous chapter, the determination of confidence limits provides, as a by-product, a significance test. A test is significant if μ_0 (null hypothesis) falls outside the confidence interval. Many statisticians prefer to make significance tests by calculating confidence intervals.

Exercises

7.3   The yields of butterfat produced (in pounds) by a random sample of ten cows are

$$65, 39, 28, 36, 50, 34, 46, 36, 54, 52.$$

Find the sample mean and estimate its standard error. Test the null hypothesis at the 5% level of significance that the population mean is 40 lb. Estimate the 95% confidence interval for the mean.

7.4   For a random sample of sixteen plantings of peaches on a particular soil type the yields (in tons/acre) were

$$7\cdot4, 1\cdot8, 2\cdot3, 2\cdot7, 10\cdot8, 5\cdot4, 6\cdot9, 7\cdot5, 1\cdot4, 9\cdot4, 6\cdot5, 3\cdot4, 9\cdot1, 6\cdot2, 8\cdot4, 6\cdot8.$$

What is an estimate of the standard error of the mean?

Test the null hypothesis at the 5% level of significance that the soil type population mean is 8·0.

What is the 99% confidence interval for the mean?

The following data are given:

$$\sum X = 96\cdot0, \quad (\sum X)^2 = 9216\cdot00, \quad \sum X^2 = 702\cdot42.$$

7.5 For the sample 13·1, 12·7, 13·3, 12·9, 13·0 from a normal distribution, find the 95% confidence interval for the mean.

7.6 Given that X is normally distributed and given the statistics $\bar{x} = 42$, $s = 5$ calculated from a sample of size 20,

(a) test the hypothesis that $\mu = 40$ (use the 5% level of significance),

(b) find the 98% confidence limits of the mean.

7.4 Comparison of Means of Two Samples

Given two independent samples of n_1 and n_2 members with means $\bar{x}_{1.}$ and $\bar{x}_{2.}$ respectively, the t distribution may be used to decide whether the two observed means differ significantly, i.e. whether the two samples may be regarded as drawn from the same normal population. This is equivalent to testing the equality of the means of the two normal homoscedastic* populations from which the samples have been drawn. Thus a test of the null hypothesis, $H_0 : \mu_1 = \mu_2 (=\mu)$, on the assumption that both populations are normal (or, provided that both samples are large, approximately normal) and have the same variance σ^2, will now be developed.

This test will be used on the many occasions when experimenters wish to compare two treatments and have a number of independent measurements on each; for example, when an animal nutritionist is studying the coefficients of digestibility of two types of feed. Here, the numbers of animals for each type might be respectively 7 and 6. In the algebra which follows X_{1j} may be considered to be the measurements made on the first type, and X_{2j} those on the second. Also for this experiment n_1 is 7 and n_2, 6.

Let $X_{1j} (j = 1, 2, \ldots, n_1)$ and $X_{2j} (j = 1, 2, \ldots, n_2)$ be the values of the variable in the first and second sample respectively.

* Homoscedastic means having equal variances.

Each of the samples provides an estimate of the population variance:

$$s_1^2 = \sum (X_{1j} - \bar{x}_1.)^2/(n_1 - 1) \quad \text{where } \bar{x}_1. = \sum_{j=1}^{n_1} X_{1j}/n_1,$$

$$s_2^2 = \sum (X_{2j} - \bar{x}_2.)^2/(n_2 - 1) \quad \text{where } \bar{x}_2. = \sum_{j=1}^{n_2} X_{2j}/n_2.$$

A weighted mean of these two estimates, the weights being their respective degrees of freedom, gives an unbiased estimate of the population variance (σ^2), i.e.

$$\begin{aligned} s^2 &= \frac{(n_1 - 1)s_1^2 + (n_2 - 1)s_2^2}{(n_1 - 1) + (n_2 - 1)} \\[2mm] &= \frac{\sum (X_{1j} - \bar{x}_1.)^2 + \sum (X_{2j} - \bar{x}_2.)^2}{n_1 + n_2 - 2} \\[2mm] &= \frac{\text{Within sum of squares*}}{\text{Within degrees of freedom}} \\[2mm] &= \text{Within mean square.} \end{aligned}$$

In addition to being unbiased, s^2 is the most efficient estimator of σ^2.

Now, by virtue of the hypothesis, $\bar{x}_1.$ and $\bar{x}_2.$ are normally distributed about the population mean (μ) with variances σ^2/n_1 and σ^2/n_2 respectively.

Therefore, since the samples are independent, the difference $\bar{x}_1. - \bar{x}_2.$ is normally distributed with mean zero and variance

$$\sigma^2 \left[\frac{1}{n_1} + \frac{1}{n_2} \right].$$

The t test which is made is

$$t = \frac{(\bar{x}_1. - \bar{x}_2.) - 0}{s\sqrt{\left(\dfrac{1}{n_1} + \dfrac{1}{n_2}\right)}} = \frac{\bar{x}_1. - \bar{x}_2.}{s\sqrt{\left(\dfrac{1}{n_1} + \dfrac{1}{n_2}\right)}}. \qquad (7.2)$$

The t table (Table III) is entered with $v = n_1 + n_2 - 2$ degrees of

---

* 'Within sum of squares' is an abbreviation for 'the sum of squares of deviations within the treatments'.

freedom. If the observed t is $< -t_{\frac{1}{2}\alpha}$ or $> t_{\frac{1}{2}\alpha}$, the null hypothesis is rejected.

The mathematical model being studied is

$$X_{ij} = \mu_i + \varepsilon_{ij} \quad (i = 1, 2; j = 1, \ldots, n_i)$$

where ε_{ij} are normally independently distributed with mean zero and variance σ^2, and the above argument is used to make the following test of significance:

Null hypothesis: $H_0 : \mu_1 = \mu_2 (= \mu)$,
Alternative hypothesis: $H_1 : \mu_1 \neq \mu_2$.

In this case, if the null hypothesis is rejected it is concluded that μ_1 is different from μ_2. If it is desired to conclude that μ_1 is greater than μ_2, the following test of significance would be carried out:

$$H_0 : \mu_1 = \mu_2,$$

$$H_1 : \mu_1 > \mu_2.$$

If $\bar{x}_{1.} < \bar{x}_{2.}$, there is no evidence to support the alternative hypothesis. If $\bar{x}_{1.} > \bar{x}_{2.}$, a t test is made. It is necessary to remember here that Table III has been prepared for two-tailed tests of significance. In a one-tailed t test as this is, only a single critical point t_α has to be found. The point t_α has to be determined so that $P(t \geqslant t_\alpha)$ is α; that is, the area under the t curve and beyond t_α has to be α. The point t_α for the one-tailed t test is found by entering Table III in the column headed 2α. For example, in a one-tailed t test with $\nu = 8$ and a significance level $\alpha = 0.05$, the critical point $t_{0.05}$ is 1.86. The number 1.86 is read in the 10% (or 2α) column of Table III. Because of the symmetry of the t distribution, $P(t \geqslant 1.86)$ is 0.05 when $P\{(t \leqslant -1.86)$ or $(t \geqslant 1.86)\}$ is 0.10.

A one-tailed test of this type would be made by a plant-breeder who had bred a new strain and was interested in it only if it out-yielded the standard variety.

The above arguments could be varied if it is desired to test the hypothesis that the samples were drawn from different normal populations with means μ_1 and μ_2 respectively but the same variances.

The following test of significance could be made.

Null hypothesis: $H_0 : \mu_1 - \mu_2 = \mu_d$,
Alternative hypothesis: $H_1 : \mu_1 - \mu_2 \neq \mu_d$.

In this case $\bar{x}_1. - \mu_1$ and $\bar{x}_2. - \mu_2$ are each normally distributed with means zero and variances σ^2/n_1 and σ^2/n_2 respectively. Hence their difference is normally distributed with mean zero and variance

$$\sigma^2\left(\frac{1}{n_1}+\frac{1}{n_2}\right).$$

Thus the t test becomes

$$t = \frac{[(\bar{x}_1. - \mu_1) - (\bar{x}_2. - \mu_2)] - 0}{s\sqrt{\left(\dfrac{1}{n_1}+\dfrac{1}{n_2}\right)}}$$

$$= \frac{(\bar{x}_1. - \bar{x}_2.) - \mu_d}{s\sqrt{\left(\dfrac{1}{n_1}+\dfrac{1}{n_2}\right)}}. \tag{7.3}$$

This is a two-tailed test. However, if the plant-breeder were interested in his new strain only if the increase in yield were significantly greater than, say, 2 bushels/acre, a one-tailed test of the following form would be made where $\mu_d = 2{\cdot}0$.

$$H_0: \mu_1 - \mu_2 = \mu_d,$$
$$H_1: \mu_1 - \mu_2 > \mu_d.$$

The observed t, as given in equation (7.3), would be calculated only if

$$\bar{x}_1. - \bar{x}_2. > \mu_d.$$

Formulae (7.2) and (7.3) can be simplified if the number of observations in each sample is the same, i.e. if $n_1 = n_2 = n$. They then become

$$t = \frac{\bar{x}_1. - \bar{x}_2.}{s\sqrt{(2/n)}}$$

and

$$t = \frac{(\bar{x}_1. - \bar{x}_2.) - \mu_d}{s\sqrt{(2/n)}}.$$

7.5 Confidence Interval for Difference between Two Means

Just as the t distribution can be used to obtain a confidence interval for the population mean when the sample mean has been found, equation (7.3) shows that the distribution can also be used to obtain a confidence interval for the difference between two means. Consideration of this equation shows that the $(1-\alpha)$ confidence interval for the difference $(\mu_1 - \mu_2)$ between two population means μ_1 and μ_2, when $\bar{x}_{1.}$ and $\bar{x}_{2.}$ are estimates of μ_1 and μ_2 from random samples, is

$$(\bar{x}_{1.} - \bar{x}_{2.}) - t_{\frac{1}{2}\alpha, v} s_{\bar{d}} \leqslant \mu_d \leqslant (\bar{x}_{1.} - \bar{x}_{2.}) + t_{\frac{1}{2}\alpha, v} s_{\bar{d}}$$

where

> $\pm t_{\frac{1}{2}\alpha, v}$ are the critical points of the t distribution with v degrees of freedom,
>
> v is the number of degrees of freedom for the estimate of the standard error of the difference,
>
> $s_{\bar{d}}$ is the standard error of the difference of the means obtained from either of the following:

(i) $s_{\bar{d}} = s \sqrt{\left(\dfrac{1}{n_1} + \dfrac{1}{n_2} \right)}$

where

$$s^2 = \frac{\sum\limits_{j=1}^{n_1} (X_{1j} - \bar{x}_{1.})^2 + \sum\limits_{j=1}^{n_2} (X_{2j} - \bar{x}_{2.})^2}{(n_1 + n_2 - 2)}$$

(ii) $s_{\bar{d}} = \dfrac{s\sqrt{2}}{\sqrt{n}}$

where

$$s^2 = \frac{\sum\limits_{i=1}^{2} \sum\limits_{j=1}^{n} (X_{ij} - \bar{x}_{i.})^2}{2n - 2}.$$

Example 7.2 The following table gives the yields in plot tests of two varieties of wheat on twelve different stations. On the assumption that the stations are not the same for the two varieties, examine the data and decide whether or not the two varieties are essentially different in yielding ability.

Variety A	22·9	19·8	24·4	27·9	23·1	25·7
	28·2	25·6	26·2	28·7	31·5	37·0

Variety B	13·7	18·2	17·5	15·1	21·6	19·2
	21·6	24·8	25·2	27·8	25·2	34·0

Let μ_1 and μ_2 be the population means for varieties A and B respectively.

$$H_0 : \mu_1 = \mu_2 \ (\sigma \text{ unknown}),$$
$$H_1 : \mu_1 \neq \mu_2,$$

Level of significance: $\alpha = 0·01$.

Since the standard deviation of the population (or populations) from which the samples were drawn is not known, it will have to be assumed that the standard deviation is the same (if the t test is to be used) and an estimate of it made from the data of the samples.

Since the variance is estimated from the samples the z distribution cannot be used, and the null hypothesis is tested using the t test in the form

$$t = \frac{\bar{x}_{1.} - \bar{x}_{2.}}{s\sqrt{2/\sqrt{n}}}.$$

The calculations necessary are

Variety A	Variety B
$n_1 = 12$	$n_2 = 12$
$\sum X_{1j} = 321·0$	$\sum X_{2j} = 263·9$
$\bar{x}_{1.} = 26·75$	$\bar{x}_{2.} = 21·99$
$\sum X_{1j}^2 = 8806·30$	$\sum X_{2j}^2 = 6168·91$
$(\sum X_{1j})^2/n_1 = 8586·75$	$(\sum X_{2j})^2/n_2 = 5803·60$
$\sum (X_{1j} - \bar{x}_{1.})^2 = 219·55$	$\sum (X_{2j} - \bar{x}_{2.})^2 = 365·31$

$$s^2 = \frac{\text{Within sum of squares}}{\text{Within degrees of freedom}}$$

$$= \frac{219·55 + 365·31}{22}$$

$$= 26·58$$

Estimated standard deviation per plot (s) $= 5 \cdot 153$,

Standard error of a variety mean $= s/\sqrt{n}$

$\qquad\qquad\qquad\qquad\qquad\qquad\qquad = 5 \cdot 153/\sqrt{12}$,

Standard error of difference of two means $= \dfrac{5.153 \times \sqrt{2}}{\sqrt{12}}$.

Then

$$t = \frac{\bar{x}_{1.} - \bar{x}_{2.}}{s\sqrt{(2/n)}}$$

$$= \frac{(26 \cdot 75 - 21 \cdot 99) \times 3 \cdot 464}{5 \cdot 153 \times 1 \cdot 414}$$

$$= 2 \cdot 263.$$

On entering the t table with $v = 22$, the 1% critical points are found to be $-2 \cdot 819$ and $2 \cdot 819$. Since the observed t is not in the critical regions, the null hypothesis is not rejected. At the 1% level, a difference in yielding ability has not been established.

In this test a 1% level of significance was chosen and the difference found to be not significant at this level. However, had a 5% level been used, the observed difference would be significant. It is important in making tests of significance that the level of significance be determined prior to the calculations. In section 1.3 the choice of the level of significance was stated to be step 2 in the four-step test of significance.

[If no calculating machine were available to make the above calculations, it would be necessary to code the data by subtracting, say, 20 from each yield and working in units of U where

$$X_{ij} = a + U_{ij} \quad (i = 1, 2; j = 1, 2, \ldots, 12)$$
$$= 20 + U_{ij};$$

and then

$$\bar{x}_{1.} = 20 + \bar{u}_{1.},$$
$$\bar{x}_{2.} = 20 + \bar{u}_{2.},$$

so that

$$\bar{x}_{1.} - \bar{x}_{2.} = \bar{u}_{1.} - \bar{u}_{2.}.$$

Also

$$\sum (X_{1j} - \bar{x}_{1.})^2 = \sum (U_{1j} - \bar{u}_{1.})^2,$$

$$\sum (X_{2j} - \bar{x}_{2.})^2 = \sum (U_{2j} - \bar{u}_{2.})^2.]$$

Exercises

7.7 A random sample of size 16 from a normal population showed a mean of 20·4 and a sum of squares of deviations from this mean equal to 135. Test (at the 1% level) the hypothesis that the population mean is 22·5 and show that the 95% confidence interval for the mean is 18·8 to 22·0.

Another sample of size 11 from a normal population (with unknown mean and variance) has a mean of 22·0 and a sum of squares of deviations from this mean equal to 90. Show that the two samples may be regarded as having been drawn from the same population.

7.8 The following data on the weight of bone in the carcases from a pre-treatment group of heifers and in the carcases of heifers which died from undernutrition are taken from Morris (1968).

	Weight of bone (kg)
Pre-treatment group	21·4, 24·2, 21·6, 22·3, 22·0, 24·9, 27·9, 22·5
Dead heifers	28·3, 30·5, 23·5, 26·8

Assuming the data fit the mathematical model

$$X_{ij} = \mu_i + \varepsilon_{ij} \quad (i = 1, 2, \ldots, t; j = 1, 2, \ldots, n_i)$$

where ε_{ij} are normally and independently distributed with zero mean and variance σ^2, test the null hypothesis (at the 5% level) that $\mu_1 = \mu_2$.

The following data should be used:

$$n_1 = 8; \quad \sum X_{1j} = 186\cdot8; \quad \sum X_{1j}^2 = 4396\cdot12;$$
$$n_2 = 4; \quad \sum X_{2j} = 109\cdot1; \quad \sum X_{2j}^2 = 3001\cdot63.$$

7.9 Consider the model

$$X_{ij} = \mu_i + \varepsilon_{ij}, \quad (i = 1, 2, \ldots, t; j = 1, 2, \ldots, n_i)$$

where ε_{ij} are normally and independently distributed with zero mean and variance σ^2. Find an estimate of σ^2.

In other words, if there are independent samples from t populations each of which has the same variance, what is an estimate of this common variance?

7.6   Method of Paired Comparisons

In some experimental situations there are likely to be large major variations between the subjects of the experiment, e.g. in a glass-house trial the variation within the glasshouse might be such as to cause a difference in yields as great as that caused by the application of fertilizer treatments. Again, a group of animals used in a nutrition experiment could give widely different results on account of differences in age, weight and sex, etc.

In such cases, if the t test is to be used to justify the inference that a particular factor is the cause of a significant change in the mean, or if the t distribution is to be used to estimate confidence intervals, it is necessary to be reasonably certain that extraneous factors are not the cause of the difference. In these situations, it is good experimental design to choose the experimental units in pairs, so that the members of each pair agree as closely as possible with respect to the factors (such as sex, age, glasshouse lighting, soil fertility, etc.) which it is desired to eliminate. After this, one member of each pair is allotted at random to the treatment being tested and the other is allotted to the control group (or second treatment). Thus two groups are formed each consisting of one individual from each pair. One group is subjected to the treatment which is being tested, and the other group is not treated. (This latter group is commonly called the *control group*.)

It would also be possible for one individual to be the subject of each of two treatments and so provide a pair of readings, e.g. the two halves of a leaf in experiments in plant pathology or the two sides of an animal in experiments in animal husbandry.

Suppose that the observations for the first treatment are

$$X_{1j}(j = 1, 2, \ldots, n)$$

and that for the second (or control) they are

$$X_{2j}(j = 1, 2, \ldots, n)$$

and let $D_j = X_{1j} - X_{2j}$. Because of the pairing (or blocking) there is the same number (n) of members in each of the two groups.

The model being considered is

$$D_j = X_{1j} - X_{2j}$$
$$= (\mu_1 + \beta_j + \varepsilon_{1j}) - (\mu_2 + \beta_j + \varepsilon_{2j})$$

$$= (\mu_1 - \mu_2) + (\varepsilon_{1j} - \varepsilon_{2j}), \quad (j = 1, 2, \ldots, n)$$
$$= \mu_d + \varepsilon_j,$$

where μ_1 and μ_2 are the population treatment means and $\mu_1 - \mu_2 = \mu_d$;

β_j are the pair (or block) effects;

ε_j are normally and independently distributed with zero mean and variance σ_d^2.

Thus $\bar{d}_. = \sum D_j/n$ is normally distributed with mean μ_d and variance σ_d^2/n. The null hypothesis being tested is $\mu_d = 0$. The t test is applied in the form

$$t = \frac{\bar{d}_.}{s_d/\sqrt{n}}$$

$$= \frac{\bar{d}_.}{s_{\bar{d}}} \tag{7.4}$$

where $s_d^2 = \sum (D_j - \bar{d}_.)^2/(n-1)$. Since the n differences D_j are used to calculate s_d^2, t has $(n-1)$ degrees of freedom. As in all tests of significance the critical regions are determined by the alternative hypothesis.

Similarly to section 7.4, if the null hypothesis is not that there is no difference in the treatment means but that $\mu_d = \delta$ (where $\delta \neq 0$) then the t test becomes

$$t = \frac{\bar{d}_. - \delta}{s_{\bar{d}}}.$$

The difference between the tests in this section and those in section 7.4 is the method of calculating the standard error of the difference of the two treatment means. In section 7.4 the variation between the observations in each of the two treatments (the within treatments sum of squares) formed the basis of the calculation. Here the variation in the differences between pairs of observations is used. In the former the variance of the difference in the two means is determined using Theorem 3.5:

$$\text{var}(\bar{x}_{1.} - \bar{x}_{2.}) = \text{var}(\bar{x}_{1.}) + \text{var}(\bar{x}_{2.})$$
$$= \frac{\sigma^2}{n} + \frac{\sigma^2}{n} = \frac{2\sigma^2}{n}.$$

An estimate of the variance of the difference in means is twice the within mean square divided by n.

However, for problems of the type considered in this section, Theorem 3.4 applies

$$\mathrm{var}(\bar{x}_{1.} - \bar{x}_{2.}) = \mathrm{var}(\bar{x}_{1.}) + \mathrm{var}(\bar{x}_{2.}) - 2\,\mathrm{cov}(\bar{x}_{1.}, \bar{x}_{2.}).$$

In general, the covariance is not zero because the two treatments have been applied to 'paired' experimental units. Rather than calculate the covariance, the variance of $(\bar{x}_{1.} - \bar{x}_{2.})$ is estimated by using the fact that the difference in the two treatment means is the mean of the n differences, i.e. $\bar{x}_{1.} - \bar{x}_{2.} = \bar{d}_{.}$. An estimate of the variance of the differences is $\sum (D_j - \bar{d}_{.})^2/(n-1)$. The variance of the mean difference is obtained by dividing this by n.

In general, $\sum (D_j - \bar{d}_{.})^2/(n-1)$ is smaller than twice the 'within mean square'. How much smaller depends on the covariance between observations made on the same pair. If measurements on the same pair tend to be similar, the covariance will be large and an efficient design will result.

However, when contemplating a 'paired' design consideration should be given to the degrees of freedom of the resultant t test. With the unpaired t test the degrees of freedom are $2n-2$, while the paired t test has $n-1$ degrees of freedom. This halving of the degrees of freedom somewhat offsets the gain which results from taking account of the covariance. Nevertheless, in general, the 'paired' design has been found to be more efficient than the 'unpaired' design.

Example 7.3 If, in Example 7.2, it had been known that the two varieties were grown on pairs of plots on each of twelve stations. the t test which should have been used is that given by equation (7.4).

It should be noted that in the following table every difference is positive. This is fairly substantial evidence that the mean of variety A exceeds the mean of variety B. If the two varietal population means were the same, half the differences would be expected to be positive and half, negative.

$$\begin{array}{lll}
\text{Null hypothesis:} & H_0 : \mu_1 = \mu_2, \\
\text{Alternative hypothesis:} & H_1 : \mu_1 \neq \mu_2. \\
\text{Model:} & X_{ij} = \mu_i + \beta_j + \varepsilon_{ij}.
\end{array}$$

Station	1	2	3	4	5	6
Variety A	22·9	19·8	24·4	27·9	23·1	25·7
Variety B	13·7	18·2	17·5	15·1	21·6	19·2
Total	36·6	38·0	41·9	43·0	44·7	44·9
Difference	9·2	1·6	6·9	12·8	1·5	6·5
Station	7	8	9	10	11	12
Variety A	28·2	25·6	26·2	28·7	31·5	37·0
Variety B	21·6	24·8	25·2	27·8	25·2	34·0
Total	49·8	50·4	51·4	56·5	56·7	71·0
Difference	6·6	0·8	1·0	0·9	6·3	3·0

$$n = 12,$$
$$\sum D_j = 57\cdot1, \qquad \bar{d}_. = 4\cdot75,$$
$$\sum D_j^2 = 437\cdot85,$$
$$(\sum D_j)^2/12 = 271\cdot70,$$
$$\sum (D_j - \bar{d}_.)^2 = 166\cdot15,$$
$$s_d^2 = \sum (D_j - \bar{d}_.)^2/11 = 15\cdot105,$$
$$t = \frac{\bar{d}_.}{s_d/\sqrt{n}} = \frac{4\cdot75}{3\cdot88/\sqrt{12}} = 4\cdot24.$$

This t has eleven degrees of freedom. If the 1% level of significance is used, the critical points are $-2\cdot201$ and $2\cdot201$. Since the observed t is greater than $2\cdot201$, the null hypothesis is rejected and it is concluded that the two varieties are different in yielding ability.

The difference in the conclusions reached in Examples 7.2 and 7.3 should be noted. These examples show clearly the need to make use of all available information about the data to be analyzed. The design of the experiment determines the method of analysis of the results. In the preceding analysis the very marked station effect (totals ranging from 36·6 to 71·0) has been removed from the estimate of error variance.

If it were desired to estimate a confidence interval for the difference in the two varietal population means (as is more likely the case in an experiment of this type), the confidence interval (say, 95%) is estimated as follows:

The 95% confidence interval is

$$\{(\bar{d}. - t_{0.05, \; n-1} \, s_d/\sqrt{n}) \leqslant \mu_d \leqslant (\bar{d}. + t_{0.05, \, n-1} \, s_d/\sqrt{n})\}.$$

Now

$$s_d/\sqrt{n} = 3{\cdot}88/\sqrt{12} = 1{\cdot}122.$$

Hence

the lower limit is $4{\cdot}75 - 2{\cdot}201 \times 1{\cdot}122 = 2{\cdot}28$,

the upper limit is $4{\cdot}75 + 2{\cdot}201 \times 1{\cdot}122 = 7{\cdot}22$,

the 95% confidence interval is $2{\cdot}28$ to $7{\cdot}22$.

Example 7.4 In an experiment to test the effect of an additional nutrient on the yielding ability of tomato plants, ten plants which acted as the control were given the standard treatment (treatment B) while ten were given the standard treatment together with the additional nutrient (treatment A). The twenty tomato plants were grown in a glasshouse in which the lighting conditions were far from uniform so that it was considered desirable to design the experiment in such a way that each treatment had plants in corresponding positions relative to the light gradient, e.g.

Light gradient $\longrightarrow$

Position	1	2	3	4	5	6	7	8	9	10
Treatment	A	B	B	A	A	A	B	A	B	A
allocation	B	A	A	B	B	B	A	B	A	B

Thus the model for the observed data is

$$X_{ij} = \mu_i + \beta_j + \varepsilon_{ij} \qquad (i = 1, 2; j = 1, 2, \ldots, 10)$$

where μ_i is the ith treatment mean and β_j is the effect of the jth position.

The yields (kilograms) of ripe fruit and the calculations used in the example are given in Table 7.1.

Test whether the observed difference is significant at the 5% level. Find the 99% confidence interval for $\mu_d (= \mu_1 - \mu_2)$ and use this interval to determine whether the observed difference is significant at 1% level.

The following test, which is *not* the correct one to use, is presented for comparison only with the correct test. This incorrect test which is about to be presented would be the appropriate one if the ten plants

TABLE 7.1 *Experimental yields of tomatoes (Example 7.4)*

Position	Treatment		Difference D_j	Position Total
	A	B		
1	0·784	0·800	−0·016	1·584
2	1·042	0·830	0·212	1·872
3	1·064	0·980	0·084	2·044
4	1·200	0·750	0·450	1·950
5	1·365	1·023	0·342	2·388
6	1·226	1·247	−0·021	2·473
7	1·401	1·212	0·189	2·613
8	1·533	1·117	0·416	2·650
9	1·651	1·505	0·146	3·156
10	1·784	1·656	0·128	3·440

$\sum_j X_{ij}$	=	13·050	11·120	1·930 $= \sum D_j$
$\bar{x}_{i.}$	=	1·305	1·112	0·193 $= \bar{d}_.$
$\sum_j X_{ij}^2$	=	17·860164	13·177332	0·618638 $= \sum D_j^2$
$(\sum X_{ij})^2/n$	=	17·03025	12·36544	0·372490 $= (\sum D_j)^2/n$
$\sum_j (X_{ij} - \bar{x}_{i.})^2$	=	0·829914	0·811892	0·246148 $= \sum (D_j - \bar{d}_.)^2$
				0·027350 $= s_d^2$
				0·0523 $\quad= s_d/\sqrt{n}$

of each treatment had been scattered at random in the glasshouse, and no attempt made to design the experiment so that the treatment difference could be studied after removal of the effect of the light gradient.

$$H_0 : \mu_1 = \mu_2,$$
$$H_1 : \mu_1 \neq \mu_2.$$

$$t = \frac{\bar{x}_{1.} - \bar{x}_{2.}}{(s\sqrt{2})/\sqrt{n}}$$

where

$$s^2 = \left[\sum_{i=1}^{2} \sum_{j=1}^{10} (X_{ij} - \bar{x}_{i.})^2 \right]/(2n-2)$$

$$= \left[\sum_{j=1}^{10} (X_{1j} - \bar{x}_{1.})^2 + \sum_{j=1}^{10} (X_{2j} - \bar{x}_{2.})^2 \right]/18$$

$$= (0·829914 + 0·811892)/18$$

$$= 1·641806/18$$

$$= 0·091211.$$

Hence

$$s = 0.3020$$

$$t = \frac{1.305 - 1.112}{(0.3020 \times 1.4142)/3.1623}$$

$$= \frac{0.193}{0.135} = 1.43.$$

Since the estimate of variance has 18 degrees of freedom, t has 18 degrees of freedom and the 5% critical points are -2.101 and 2.101. The observed t is not in the critical region and the null hypothesis is not rejected at the 5% level of significance. It is concluded that the difference between the yields of the two treatments can be accounted for by random variation.

The above test has been based on the assumption that the model

$$X_{ij} = \mu_i + \varepsilon_{ij}$$

is appropriate. *This is not correct* and the appropriate test to use here is that of the method of paired comparisons since the model is

$$X_{ij} = \mu_i + \beta_j + \varepsilon_{ij}.$$

For this method, the calculations lead to

$$t = \frac{\bar{d}_.}{\text{s.e. of } \bar{d}_.}$$

$$= \frac{\bar{d}_.}{[\sum (D_j - \bar{d}_.)^2/(n-1)]^{\frac{1}{2}}/\sqrt{n}}$$

$$= \frac{0.193}{0.0523} = 3.69.$$

For nine degrees of freedom, this is significant at the 1% level. The conclusion to be drawn from this test, *unlike that of the previous incorrect test*, is that there is a difference in the yields of the two treatments.

In the first (incorrect) test no account was taken of the position of each plant in the glasshouse, while in the latter test comparisons were made only between the plants in like positions. This is the reason for the non-significance in the first case and significance when the test was carried out correctly. The marked positional effect is obvious from the position totals which increase from 1·584 to 3·440.

Exercises

7.10   A group of 7 eight-week old chickens reared on a high protein diet weighed 13, 12, 15, 17, 14, 17, 15 ozs. The weights of a second group of 5 chickens similarly treated except that they received a low protein diet were 9, 15, 14, 11, 11 ozs.

Use the t test to test whether there is significant evidence at the 5% level that additional protein has increased the weight of the chickens.

[Use $H_0: \mu_1 = \mu_2; H_1: \mu_1 > \mu_2.$]

7.11   The following table gives the weights of chickens at six weeks of age for two samples A and B of nine chickens each. In series A the chickens were reared on open range and in B in confinement. Use the data to test at the 5% level whether the different methods of maintenance affect the weights of chickens. Assume that the data fit the model

$$X_{ij} = \mu_i + \varepsilon_{ij}$$

where the ε_{ij} are independently and normally distributed with zero means.

	Weight (oz) (X_{ij})								
Series A ($i = 1$)	10	10	13	12	16	9	12	11	15
Series B ($i = 2$)	15	11	16	16	17	10	14	12	15

If it were known that the chickens in series B were the offspring of the same nine parents as those in series A, in the order shown, what difference would this make to the test carried out? Apply the appropriate test (at the 5% level) if this knowledge were available. Find the 95% confidence interval for the difference in weights in both cases.

7.12   A random sample of size nine is chosen from a population of grasshoppers. The weights of the grasshoppers are normally distributed with unknown mean and unknown standard deviation.

An unbiased estimate (s^2) of the population variance provided by the sample is $0·09(g)^2$. What is the probability that the sample mean $(\bar{x})$ exceeds the population mean by $0·29$ g? (Compare the answer for this exercise with that for Exercise 6.5.)

7.13   Two wheat varieties A and B have been compared in pairs of plots located on six different properties in a given region. The following grain yields (in bushels per plot) were obtained:

Property	1	2	3	4	5	6
Variety A	25	16	13	19	23	17
Variety B	18	14	9	15	18	12

The two varieties were allocated at random to the two plots on each property. Test the significance of the difference in yield shown by the two varieties by means of a t test.

7.14   For a random sample of nine pigs fed on diet A the increases in weight in a certain period were

$$10, 7, 16, 16, 13, 12, 8, 14, 15.$$

For another random sample of nine pigs fed on diet B the increases in the same period were

$$13, 20, 15, 14, 14, 18, 10, 20, 21.$$

Assume the weight gains (W) may be represented by the model

$$W_{ij} = \mu_i + \varepsilon_{ij} \quad (i = 1, 2; j = 1, \ldots, 9)$$

where ε_{ij} are distributed independently with common variance.
For the variable, weight gain, estimate
(a) the standard error of a diet mean;
(b) the standard error of the difference of the two diet means.
If μ_1 is the population mean for diet A, and μ_2 is the population mean for diet B, test the hypothesis that $\mu_1 = \mu_2$.
Find the 95% confidence limits for the difference between the two diet means.

7.15   The set of data used in this exercise is part of the results of an experiment reported by Holford (1968). The experiment was conducted on ten wheat-growing soils in pot culture during 1965–66.

Table 7.2 shows the effect of phosphorus and sulphur on yield of dry matter and nitrogen by wheat following lucerne, and nitrogen fixation by lucerne on the ten soils.

TABLE 7.2 *Experimental data for Exercise 7.15*

	Means ($\bar{y}_{ij.}$) of three pots								
Soil	Yield of dry matter $-$PS $+$PS Difference			Yield of nitrogen $-$PS $+$PS Difference			Nitrogen fixation $-$PS $+$PS Difference		
1	5·54	6·80	1·26	59	69	10	261	459	197
2	6·52	7·55	1·03	66	90	24	407	756	349
3	6·19	7·24	1·05	66	81	15	264	729	465
4	6·47	9·36	2·89	68	101	33	232	896	664
5	6·46	7·48	1·02	71	83	12	204	721	517
6	5·52	6·96	1·44	64	84	20	221	610	389
7	5·93	7·76	1·83	58	84	26	430	737	307
8	6·70	8·63	1·93	72	97	25	576	1338	762
9	6·24	7·49	1·25	67	85	18	429	1200	771
10	6·10	6·75	0·65	66	75	9	210	604	394

Assume the data for each of the three variables satisfy the statistical model

$$\bar{y}_{ij.} = \mu_i + \beta_j + \varepsilon_{ij} \qquad (i = 1, 2; j = 1, 2, \dots, 10)$$

and

$$D_j = \bar{y}_{1j.} - \bar{y}_{2j.}$$

$$= (\mu_1 - \mu_2) + \varepsilon_j \qquad (\varepsilon_j = \varepsilon_{1j} - \varepsilon_{2j})$$

where μ_1 and μ_2 are the treatment means,

β_j is the effect of the *j*th soil,

ε_j are error terms, normally and independently distributed with zero mean and variance σ_d^2.

Test at the 5% level the hypothesis

$$H_0 : \mu_1 = \mu_2,$$

$$H_1 : \mu_1 \neq \mu_2.$$

REFERENCES

HOLFORD, I. C. R. (1968). 'Nitrogen Fixation by Lucerne on Several Soils in Pot Culture', *Aust. J. Exper. Agric. and Animal Husb.*, **8**, 555–560.

MORRIS, J. G. (1968). 'Feeding of Heifers for Survival on Whole and Cracked Sorghum Grain Under Simulated Drought Conditions', *Aust. J. Exper. Agric. and Animal Husb.*, **8**, 668–673.

COLLATERAL READING

BROWNLEE, K. A. (1965). *Statistical Theory and Methodology* (2nd edition). John Wiley, New York. Chapter 9.

FINNEY, D. J. (1964). *An Introduction to Statistical Science in Agriculture* (2nd edition). Oliver & Boyd, Edinburgh; John Wiley, New York. Chapter 5.

GOULDEN, C. H. (1952). *Methods of Statistical Analysis* (2nd edition). John Wiley, New York. Chapter 4.

LI, J. C. R. (1964). *Statistical Inference*, Vol. I. Edwards Brothers, Ann Arbor, Michigan. Chapter 8.

STEEL, R. G. D. and Torrie, J. H. (1960). *Principles and Procedures of Statistics*. McGraw-Hill, New York. Chapter 5.

CHAPTER 8

The F Test

8.1 Ratio of Variances of Samples from Normal Populations

In comparing two samples it may be important to consider not only whether there is a significant difference between their means but also whether their variances differ significantly. For example, a comparison of the variability of yields of peaches grown on two different soil types might be required, or again, a test of whether one scientific instrument gives less variable readings than another.

The test which is used to compare two estimated variances is the F test. The null and alternative hypotheses are

$$H_0 : \sigma_1^2 = \sigma_2^2, \qquad H_1 : \sigma_1^2 \neq \sigma_2^2,$$

or

$$H_0 : \sigma_1^2 = \sigma_2^2, \qquad H_1 : \sigma_1^2 > \sigma_2^2,$$

whereas the t test is used to test the means; e.g.

$$H_0 : \mu_1 = \mu_2, \qquad H_1 : \mu_1 \neq \mu_2,$$

or

$$H_0 : \mu_1 = \mu_2, \qquad H_1 : \mu_1 > \mu_2.$$

The F test which will be developed in this chapter as a test for comparing variances will be used in the next chapter to test the hypothesis

$$H_0 : \mu_1 = \mu_2 = \mu_3 = \ldots = \mu_p$$

where the equality of $p(p \geqslant 2)$ treatment or varietal means is being examined. Thus the F test will later be used to test the null hypothesis which is an extension of that tested by the t test in the previous chapter.

At this point it might be noted that whereas the t test for equality of means assumes equality of variances, the F test for equality of

L

variances does not assume equality of means. This assumption of equality of variance is a necessary one for the t test and if this assumption does not hold, a different test for equality of means has to be made. A test known as the Behrens–Fisher test is used. Tables for this are given by Fisher and Yates (1949). An approximation presented by Cochran and Cox (1950) is sufficiently accurate for most purposes. The only assumptions underlying the F test are those indicated in the following paragraph. No assertions about the normal population means are made.

Suppose two independent random samples of size n_1 and n_2 are drawn from the same normal population or from normal populations having equal variances.

Let X_{ij} ($i = 1, 2; j = 1, 2, \ldots, n_i$) be the observation made on the jth member of the ith sample. Define

$$\bar{x}_{i.} = \left(\sum_{j=1}^{n_i} X_{ij} \right) \Big/ n_i \ .$$

Then

$$s_1^2 = \sum (X_{1j} - \bar{x}_{1.})^2 / (n_1 - 1),$$
$$s_2^2 = \sum (X_{2j} - \bar{x}_{2.})^2 / (n_2 - 1).$$

are both unbiased estimates of the unknown population variance σ^2.

Then the *variance* ratio

$$F = \frac{s_1^2}{s_2^2} = \frac{(n_2 - 1) \sum (X_{1j} - \bar{x}_{1.})^2}{(n_1 - 1) \sum (X_{2j} - \bar{x}_{2.})^2} = \frac{\sum (X_{1j} - \bar{x}_{1.})^2}{\sum (X_{2j} - \bar{x}_{2.})^2} \qquad \text{if } n_1 = n_2 \ .$$

The distribution of F has two parameters, v_1 the degrees of freedom of s_1^2, and v_2 the degrees of freedom of s_2^2. The probability curve of F (Figure 8.1) is skew and ranges from 0 to ∞. It has a mean at $v_2/(v_2 - 2)$ and a mode at $v_2(v_1 - 2)/v_1(v_2 + 2)$.

The F distribution is related to the t, χ^2 and normal distributions. If $v_1 = 1$ and v_2 is finite, the distribution of F is the same as that of t^2. Also if v_1 is finite and v_2 becomes infinite, the limiting form of the distribution of $v_1 F$ is the χ^2 distribution with v_1 degrees of freedom. Because of the relationships between the F and χ^2 distributions and between the χ^2 and normal distributions, the F distribution is also related to the normal distribution.

Table IV in the appendix gives percentage points of the F distri-

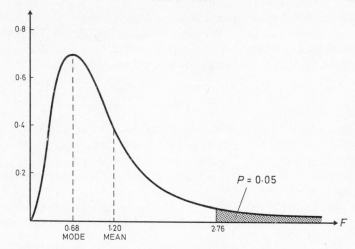

Fig. 8.1   *F* distribution with degrees of freedom 10 and 12

bution. For various probabilities (α), points F_α on the F distribution are tabulated where F_α are such that $P(F \geqslant F_\alpha) = \alpha$. On entering Table IV at the column headed 8 and the row labelled 14, it is seen that, for the F distribution with 8 and 14 degrees of freedom, the probabilities of exceeding 2·70 and 4·14 are respectively 5% and 1%.

Exercises

8.1   For the F distribution with degrees of freedom (6, 8), what is the value of F_0 such that $P(F \geqslant F_0) = 0.05$?

8.2   For a particular F distribution with degrees of freedom (v, 26), $P(F \geqslant 2.96) = 0.01$. What is the value of v?

8.3   For the t_{20} distribution, obtain the 5% and 1% points. What relationship is found between (a) $F_{0.05}$ and $t_{0.05}$ and (b) $F_{0.01}$ and $t_{0.01}$, where $F_{0.05}$ and $F_{0.01}$ are the 5% and 1% points of the F distribution with degrees of freedom 1 and 20? What is the relationship between $t_{\alpha, v}$ and $F_{\alpha, 1, v}$?

8.2   The Two-tailed *F* Test

Suppose it is desired to test whether two independent estimates of variance, s_1^2 and s_2^2 based on v_1 and v_2 degrees of freedom respectively are significantly different (v_1 and v_2 need not be equal). The null hypothesis to be tested is that the two samples have been drawn at random from normal populations of equal variance, σ^2 say.

$$H_0: \sigma_1^2 = \sigma_2^2 \quad (= \sigma^2),$$
$$H_1: \sigma_1^2 \neq \sigma_2^2.$$

Because of the alternative hypothesis this is a two-tailed test of significance. However, generally only the upper percentage points of the F distribution are tabulated and this necessitates some interpretation of the F table when two-tailed tests of significance are made.

Suppose that s_1^2/s_2^2 were to be taken as the test statistic and that $F_{R(v_1, v_2)}$ and $F_{L(v_1, v_2)}$ are the critical points for an α-level significance test so that

$$P\{F_{(v_1, v_2)} > F_{R(v_1, v_2)}\} = P\{F_{(v_1, v_2)} < F_{L(v_1, v_2)}\} = \tfrac{1}{2}\alpha.$$

Then if either $s_1^2/s_2^2 > F_{R(v_1, v_2)}$ or $s_1^2/s_2^2 < F_{L(v_1, v_2)}$ the null hypothesis is rejected. $F_{R(v_1, v_2)}$ can be read from tables but $F_{L(v_1, v_2)}$ is not tabulated. Thus if $s_1^2 < s_2^2$ another way of determining whether s_1^2/s_2^2 is significantly small has to be found. Since small values of s_1^2/s_2^2 are large values of s_2^2/s_1^2 ($s_1^2/s_2^2 = 0{\cdot}04$; $s_2^2/s_1^2 = 25$) the observed value of s_2^2/s_1^2 can be compared with $F_{R(v_2, v_1)}$ where $P\{F_{(v_2, v_1)} > F_{R(v_2, v_1)}\} = \tfrac{1}{2}\alpha$.

Thus the two-tailed test of significance is made by dividing the larger mean square by the smaller mean square and using either $F_{R(v_1, v_2)}$ or $F_{R(v_2, v_1)}$ whichever is appropriate. In determining $F_{R(v_1, v_2)}$ or $F_{R(v_2, v_1)}$ it is necessary to remember that these are points in the F distributions corresponding to probability $\tfrac{1}{2}\alpha$. For a 5% test of significance the upper $2\tfrac{1}{2}$% point of the distribution is used. The procedure is illustrated in Figure 8.2.

Example 8.1 A cereal chemist investigating the protein percentage of two different strains of wheat obtained the following data. Test whether the two strains differ significantly in variance.

Strain A	11·0	11·5	11·8	12·2	11·2	12·6	11·6	12·0
	12·4	11·6	12·7	12·5	12·0	12·5		
Strain B	15·4	14·5	12·5	15·1	16·6	14·6	11·5	14·3

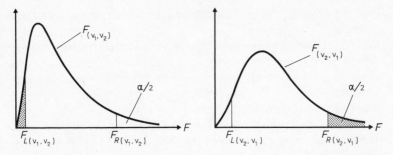

Fig. 8.2 Illustration of the two-tailed *F* test. Unshaded critical regions correspond; shaded critical regions correspond. Area associated with each critical region is $\alpha/2$

It is apparent from these results that the mean readings of the two strains are different. However, the interest here is in variability of the strains rather than in a comparison of their means.

$$H_0: \sigma_A^2 = \sigma_B^2,$$

$$H_1: \sigma_A^2 \neq \sigma_B^2,$$

Level of significance: $\alpha = 0.05$.

Let $X_{ij} (i = 1, 2; j = 1, 2, \ldots, n_i)$ be the jth observation for the ith strain.

$$
\begin{array}{ll}
n_1 = \quad 14 & n_2 = \quad 8 \\
\sum X_{1j} = 167\cdot6 & \sum X_{2j} = \quad 114\cdot5 \\
\sum X_{1j}^2 = 2010\cdot20 & \sum X_{2j}^2 = 1657\cdot13 \\
(\sum X_{1j})^2/n_1 = 2006\cdot41 & (\sum X_{2j})^2/n_2 = 1638\cdot78 \\
\sum (X_{1j} - \bar{x}_{1.})^2 = \quad 3\cdot79 & \sum (X_{2j} - \bar{x}_{2.})^2 = \quad 18\cdot35
\end{array}
$$

$$s_1^2 = [\sum (X_{1j} - \bar{x}_{1.})^2]/(n_1 - 1) = 0\cdot291,$$

$$s_2^2 = [\sum (X_{2j} - \bar{x}_{2.})^2]/(n_2 - 1) = 2\cdot621.$$

Since $s_2^2 > s_1^2$,

$$F_{7,13} = \frac{s_2^2}{s_1^2} = \frac{2\cdot621}{0\cdot291} = 9\cdot01.$$

Since the level of significance is 0·05 and since the test is a two-tailed test, the critical point in the $F_{7,13}$ distribution is 3·48. The observed F is greater than this and hence the null hypothesis is rejected. The alternative hypothesis is accepted and it is concluded that the variability in protein determinations for the two strains is different.

Notes: (a) The difference between the variances is so large that if it were desired to test the hypothesis of equality of means, the t test in the form given in the previous chapter should not be used. This is an instance when the Behrens–Fisher test, or the approximation of Cochran and Cox should be used.

(b) If it were desired to give standard errors for the strain means, these would be calculated from the data for the individual strains, e.g.

$$\text{standard error of mean of strain } A = s_1/\sqrt{n_1}$$
$$= \sqrt{\frac{0\cdot291}{14}}$$
$$= 0\cdot14.$$

8.3   Significance of Variance Ratio

Suppose it was desired to examine whether the variance of the protein percentages for strain A is less than that for strain B. The test of significance carried out is

$$H_0: \sigma_B^2 = \sigma_A^2,$$
$$H_1: \sigma_B^2 > \sigma_A^2.$$

The F statistic used is the previous one,

$$F_{7,\,13} = s_2^2/s_1^2 = 9\cdot01.$$

This is a one-tailed test, since from the acceptance of the alternate hypothesis the inference is drawn that the variance of the B determinations is greater than the variance of the A determinations (and not, as in the previous test, either 'greater than' or 'less than'). Since this is a one-tailed test, the one-tailed probabilities shown in Table IV are used. For a 5% test of significance the critical point for $F_{7,\,13}$ would be $2\cdot83$ since $P(F_{7,\,13} > 2\cdot83) = 0\cdot05$.

This is the form in which the F test will be applied in the analysis of variance, and is the general way in which the F table is used.

Exercises

8.4   For a random sample of twelve pigs fed on an improved diet the weight gains over a certain period of time were

13, 15, 14, 12, 22, 7, 18, 21, 10, 17, 23, 8 lb.

For another random sample of ten pigs fed on a control diet the weight gains over the same period were

6, 17, 12, 14, 15, 9, 8, 13, 16, 10 lb.

Use the F test to show that the estimates of the population variance from these samples are not significantly different. (This is a two-tailed test.)

8.5 The data of Table 8.1 are taken from Morris (1968) and relate to the weight of soft tissues and bone in the carcases from a pre-treatment group of heifers and in the carcases of heifers which died from undernutrition. If σ_1^2 and σ_2^2 are the population variances for the pre-treatment and dead heifers respectively, test for both 'soft tissues' and 'bone' the hypothesis that $\sigma_1^2 = \sigma_2^2$.

TABLE 8.1   Data for Exercise 8.5

	Pre-treatment group			Dead heifers	
Animal number	Soft tissues kg	Bone kg	Animal number	Soft tissues kg	Bone kg
36	63·9	21·4	65	36·0	28·3
43	75·8	24·2	45	39·6	30·5
42	72·7	21·6	48	40·6	23·5
55	67·5	22·3	49	40·4	26·8
78	65·1	22·0			
56	68·0	24·9			
80	83·8	27·9			
51	58·2	22·5			
n	8	8	n	4	4
$\sum X$	555·0	186·8	$\sum X$	156·6	109·1
$\sum X^2$	38942·08	4396·12	$\sum X^2$	6144·68	3001·63

8.6 Byth and Waite (1962) presented the following results of chemical analyses on seed produced by a range of soybean varieties in two contrasting seasons. For each of the variables, protein and oil, test the hypothesis that the variance in 1957 is less than or equal to that in 1959 against the alternative that the variance in 1957 is greater than that in 1959.

Protein percentage

1957	42·7	38·4	42·3	43·0	42·0	43·9	43·8	43·7	40·6
	43·5	46·4	36·5						
1959	43·8	36·8	33·8	36·6	39·5	40·9	42·3	37·1	

Oil percentage

1957	15·1	16·2	18·8	17·1	17·1	16·1	17·5	11·3	15·5
	18·7	13·4	16·5						
1959	16·7	18·8	19·9	21·3	17·6	17·1	18·0	22·0	

[The following data should be used

Protein: $\sum X_{1j} = 506\cdot8$; $\sum X_{1j}^2 = 21\,481\cdot30$;
$\sum X_{2j} = 310\cdot8$; $\sum X_{2j}^2 = 12\,153\cdot44$.

Oil: $\sum X_{1j} = 193\cdot3$; $\sum X_{1j}^2 = 3\,163\cdot61$;
$\sum X_{2j} = 151\cdot4$; $\sum X_{2j}^2 = 2\,892\cdot20$.]

8.7 Andrews and Hart (1962) determined the following vitamin B_{12} concentration ($\mu g/g$) in the livers and kidneys of (a) lambs treated with cobaltic oxide pellets and (b) untreated lambs grazing cobalt-deficient pastures.

Group	Liver							
Cobalt	1·60	1·50	0·83	0·80	0·79	0·77	0·77	0·73
	0·71	0·67	0·50	0·44				
Cobalt deficient	0·16	0·15	0·13	0·12	0·12	0·11	0·10	0·09

Group	Kidney							
Cobalt	0·52	0·49	0·62	0·41	0·58	0·54	0·75	0·43
	0·62	0·59	0·43	0·29				
Cobalt deficient	0·21	0·19	0·25	0·25	0·20	0·19	0·26	0·25

Examine the hypothesis that vitamin B_{12} concentrations in the livers and in the kidneys are less variable in cobalt-treated lambs than in cobalt-deficient lambs.

[The following data should be used

$$\text{Liver}: \sum X_{1j} = 10\cdot11; \quad \sum X_{1j}^2 = 9\cdot8783;$$
$$\sum X_{2j} = 0\cdot98; \quad \sum X_{2j}^2 = 0\cdot1240.$$

$$\text{Kidney}: \sum X_{1j} = 6\cdot27; \quad \sum X_{1j}^2 = 3\cdot4399;$$
$$\sum X_{2j} = 1\cdot80; \quad \sum X_{2j}^2 = 0\cdot4114.]$$

8.8 In a chemical analysis repeated measurements by two experimenters A and B gave the following results:

A 6·81, 6·80, 6·78, 6·82, 6·78, 6·73, 6·74;

B 6·64, 6·66, 6·86, 6·81, 6·76.

Assuming that an experimenter is more reliable the smaller the variance of his measurements, test whether the difference between A and B is large enough to warrant the conclusion that A is better than B.

REFERENCES

ANDREWS, E. D. and HART, L. I. (1962). 'A Comparison of Vitamin B_{12} Concentrations in Livers and Kidneys from Cobalt-Treated and Mildly Cobalt-Deficient Lambs', *N.Z.J. Agric. Res.* **5**, 403–408.

BYTH. D. E. and WAITE, R. B. (1962). 'Soybeans for Sub-tropical Queensland', *Aust. J. Exper. Agric. and Anim. Husb.*, **2**, 110–125.

COCHRAN, W. G. and COX, GERTRUDE M. (1957). *Experimental Designs*. John Wiley, New York.

FISHER, R. A. and YATES. F. (1949). *Statistical Tables for Biological, Agricultural and Medical Research* (3rd edition). Oliver & Boyd, Edinburgh.

MORRIS, J. G. (1968). 'Feeding of Heifers for Survival on Whole and Cracked Sorghum Grain Under Simulated Drought Conditions', *Aust. J. Exper. Agric. and Anim. Husb.*, **8**, 668–673.

COLLATERAL READING

LI, J. C. R. (1964). *Statistical Inference*, Vol. I. Edwards Brothers, Ann Arbor, Michigan. Chapter 9.

MODE, E. B. (1951). *Elements of Statistics*. Prentice-Hall, Englewood Cliffs, New Jersey. Chapter 14.

SNEDECOR, G. W. (1956). *Statistical Methods* (5th edition). Iowa State University Press, Ames, Iowa. Chapters 4 & 10.

CHAPTER 9

The Analysis of Variance

9.1 Introduction

In Chapters 5 to 7, methods were developed to study the difference between two means. If the variances of the populations from whence the samples have been drawn are known a z test is used, and if the variances are not known (but equal) a t test is used. However, experimentalists are frequently interested in the differences among more than two treatment means. In these cases the most commonly used technique is the analysis of variance in which the F distribution is used if a test of significance is to be made concerning the equality of the treatment means. This statistical method was introduced by Fisher to deal with problems in agricultural and biological research, but is now used in many other fields.

The analysis of variance was defined by Fisher (1948) as:

'The separation of the variance ascribable to one group of causes from the variance ascribable to other groups.'

The overall approach in the analysis of variance is to partition the total variation in the observed set of data into a number of components. These components are then used to estimate variances (or some functions of variances), and if a test of significance is to be made the estimated variances are compared by means of an F test.

9.2 One Criterion of Classification; The Completely Randomized Design

This is the simplest case—that in which the effect of a single factor is being examined. Here only the effect of the treatments is isolated from the random or unexplained variation; the total variation is partitioned into two parts—the variation due to treatments and residual variation.

The experimental design is known as a completely randomized

design because the treatments are randomly allocated to the experimental units. There is no restriction that each treatment be applied to the same number of experimental units. For this reason the design is extremely flexible and has found wide use in animal experiments. However, it is often found to be inefficient, especially in the case of field experiments where the randomized complete block design (to be considered later in this chapter) has been found to be generally very efficient.

Suppose that 60 cows are chosen at random from a large herd and that the milk yield of each over a certain period of time is recorded. If in the sample of 60 there were, say, 4 different breeds, a one-way analysis of variance could be used to test the hypothesis of equality of mean milk yield for the 4 different breeds.

The general theory of the one-way analysis of variance is now presented. Suppose that t treatments are to be tested for 'yield' in an experiment and that N experimental units are available. Suppose further that the ith treatment is applied to n_i units. (In relation to the milk problem, $N = 60$, $t = 4$ (the number of breeds) and n_1, n_2, n_3, n_4 are the number of cows of each of the 4 breeds.) Then

$$\sum n_i = N.$$

Let Y_{ij} be the yield of the jth unit of the ith treatment. Then i ranges from 1 to t; j from 1 to n_i.

The yields, treatment means and totals may be set out as in Table 9.1.

TABLE 9.1 *Data to be analyzed by one-way analysis of variance*

Treatments	Yields	Number of units	Treatment Means	Treatment Totals
1	$Y_{11} \, Y_{12} \ldots Y_{1j} \ldots Y_{1n_1}$	n_1	$\bar{y}_{1.}$	$Y_{1.}$
2	$Y_{21} \, Y_{22} \ldots Y_{2j} \ldots \ldots Y_{2n_2}$	n_2	$\bar{y}_{2.}$	$Y_{2.}$
3	$Y_{31} \, Y_{32} \ldots Y_{3j} \ldots \ldots \ldots Y_{3n_3}$	n_3	$\bar{y}_{3.}$	$Y_{3.}$
⋮				
i	$Y_{i1} \, Y_{i2} \ldots Y_{ij} \ldots Y_{in_i}$	n_i	$\bar{y}_{i.}$	$Y_{i.}$
⋮				
t	$Y_{t1} \, Y_{t2} \ldots Y_{tj} \ldots \ldots Y_{tn_t}$	n_t	$\bar{y}_{t.}$	$Y_{t.}$

The data are assumed to satisfy the following model which is an extension of that used in the case of the unpaired t test.

$$Y_{ij} = \mu + \alpha_i + \varepsilon_{ij} \qquad (i = 1, 2, \ldots, t; j = 1, 2, \ldots, n_i)$$
$$= \mu_i + \varepsilon_{ij}$$

where μ_i is the population mean for the ith treatment;

α_i is the effect of the ith treatment;

μ is the mean of the μ_i so that $\sum \alpha_i = 0$;

ε_{ij} are random error terms which are normally distributed with mean zero and variance σ^2.

The null and alternative hypotheses in the test of significance are

$$H_0: \mu_1 = \mu_2 = \mu_3 = \ldots = \mu_t,$$

H_1: at least one pair of means is different.

The null hypothesis is that the means of the t normal populations are all equal. The general method of attack is to examine whether the variability observed in the treatment means is more than would be reasonably expected under the null hypothesis that the treatment population means are equal.

Define

$$\bar{y}_{i.} = \left(\sum_j Y_{ij} \right) / n_i \qquad \text{(Treatment mean)},$$

$$\bar{y}_{..} = \left(\sum_i \sum_j Y_{ij} \right) / N \qquad \text{(General mean)},$$

$$Y_{i.} = \sum_j Y_{ij} \qquad \text{(ith Treatment total)},$$

$$Y_{..} = \sum_i \sum_j Y_{ij} = \sum_i Y_{i.} \qquad \text{(Grand total)}.$$

Now

$$Y_{ij} - \bar{y}_{..} = Y_{ij} - \bar{y}_{i.} + \bar{y}_{i.} - \bar{y}_{..},$$
$$(Y_{ij} - \bar{y}_{..})^2 = (Y_{ij} - \bar{y}_{i.})^2 + (\bar{y}_{i.} - \bar{y}_{..})^2 + 2(Y_{ij} - \bar{y}_{i.})(\bar{y}_{i.} - \bar{y}_{..}).$$

For a particular i,

$$\sum_j (Y_{ij} - \bar{y}_{..})^2 = \sum_j [(Y_{ij} - \bar{y}_{i.})^2 + (\bar{y}_{i.} - \bar{y}_{..})^2 + 2(Y_{ij} - \bar{y}_{i.})(\bar{y}_{i.} - \bar{y}_{..})]$$
$$= \sum_j (Y_{ij} - \bar{y}_{i.})^2 + n_i(\bar{y}_{i.} - \bar{y}_{..})^2 + 0.$$

[The last term is zero because

$$\sum_j (Y_{ij} - \bar{y}_{i.})(\bar{y}_{i.} - \bar{y}_{..}) = (\bar{y}_{i.} - \bar{y}_{..}) \sum_j (Y_{ij} - \bar{y}_{i.}) = (\bar{y}_{i.} - \bar{y}_{..}).0.]$$

Summing over all i,

$$\sum_i \sum_j (Y_{ij} - \bar{y}_{..})^2 = \sum_i \sum_j (Y_{ij} - \bar{y}_{i.})^2 + \sum_i n_i(\bar{y}_{i.} - \bar{y}_{..})^2$$

$$= \sum_i n_i(\bar{y}_{i.} - \bar{y}_{..})^2 + \sum_i \sum_j (Y_{ij} - \bar{y}_{i.})^2. \quad (9.1)$$

The left-hand member of equation (9.1) is the total of the squared deviations of the sample values of the variate (yield, in this case) from the general mean and is a measure of the 'total variation'.

The right-hand side of the equation shows that the total variation may be resolved into two components:

(a) $\sum_i n_i(\bar{y}_{i.} - \bar{y}_{..})^2$ is the variation which would have resulted if there had been no variation within each treatment. This may be appreciated by putting all $Y_{ij} = \bar{y}_{i.}$. Such a substitution does not alter the first term on the right-hand side of the equation but the second term is then zero. For this reason the first term is called the 'between treatments variation'.

(b) $\sum \sum (Y_{ij} - \bar{y}_{i.})^2$ is the residual 'variation within treatments'.

Briefly, equation (9.1) shows that

Total variation = Variation between treatments
                 + Variation within treatments

The sums of squares in equation (9.1) are used in the analysis of variance and are computed from the following formulae:

$$\text{Total S.S.} = \sum_i \sum_j (Y_{ij} - \bar{y}_{..})^2$$

$$= \sum_i \sum_j Y_{ij}^2 - \frac{Y_{..}^2}{N}; \quad (9.2)$$

$$\text{Between treatments S.S.} = \sum_i n_i(\bar{y}_{i.} - \bar{y}_{..})^2$$

$$= \sum_i \frac{Y_{i.}^2}{n_i} - \frac{Y_{..}^2}{N}; \quad (9.3)$$

$$\text{Within treatments S.S.} = \sum_i \sum_j (Y_{ij} - \bar{y}_{i.})^2$$

$$= \sum_i \sum_j Y_{ij}^2 - \sum_i \frac{Y_{i.}^2}{n_i}$$

$$= \sum_i \left[\sum_j Y_{ij}^2 - \frac{Y_{i.}^2}{n_i} \right]. \quad (9.4)$$

The term, $Y^2_{.}/N$, is commonly called the 'correction term'.

These formulae are analogous to that used in computing a single sample variance:

$$\sum_{i=1}^{n} (X_i - \bar{x})^2 = \sum X_i^2 - (\sum X_i)^2/n.$$

Formula (9.3) shows that the between treatments sum of squares is calculated by squaring each treatment total, dividing each square by the number of units to which the treatment has been applied, summing the t quotients and finally subtracting the correction term. Of course, if each treatment has been applied to the same number of units (n say), formula (9.3) may be simplified and the between treatments sum of squares is then obtained by squaring the treatment totals, summing the squares, dividing the sum by n and finally subtracting the correction term.

Frequently the within treatments sum of squares is obtained by subtraction, i.e.

Within treatments S.S. = Total S.S. − Between treatments S.S.

However, as a check on the mechanical operations, it is good practice to find the Within treatments S.S. separately by using formula (9.4) and checking that

Total S.S. = Between treatments S.S. + Within treatments S.S.

9.3   Estimate of Population Variance from Within Treatments Sum of Squares

Each of the t treatment groups is a random sample from a population with variance σ^2. While the means might be different it is assumed that the variances for each treatment population are the same. Thus the yields (Y_{1j}) for the first treatment might be used to estimate the variance. An unbiased estimate of the population variance σ^2 is

$$s_1^2 = \sum (Y_{1j} - \bar{y}_{1.})^2/(n_1 - 1).$$

This estimate has $(n_1 - 1)$ degrees of freedom. Similarly the yields (Y_{2j}) for the second treatment provide a second estimate based on $(n_2 - 1)$ degrees of freedom; and, in general,

$$\sum (Y_{ij} - \bar{y}_{i.})^2/(n_i - 1)$$

is an unbiased estimate (s_i^2) of the population variance based on $(n_i - 1)$ degrees of freedom.

As in the case of the unpaired t test where two estimates of variance were combined (section 7.4), the estimates from each treatment group are combined here to obtain the most efficient estimate of the population variance. This estimate s_e^2 is given by

$$s_e^2 = \frac{\sum (n_i - 1) s_i^2}{\sum (n_i - 1)}$$

$$= \sum_i \sum_j (Y_{ij} - \bar{y}_{i.})^2 / (N - t)$$

$$= \frac{\text{Within treatments sum of squares}}{\text{Within treatments degrees of freedom}}$$

$$= \text{Within treatments mean square.}$$

s_e^2 may be thought of as a weighted mean of the t estimates which may be obtained from the t treatment groups, the weights being the degrees of freedom. The within treatments mean square (s_e^2) is an unbiased estimate (based on N-t degrees of freedom) of the common population variance (σ^2). It is important to note that, provided the yields of the different treatments are distributed with equal variances, s_e^2 is unbiased even if the null hypothesis is not true and the treatment population means, μ_i, are not all equal.

9.4 Estimate of Population Variance from Between Treatments Sum of Squares

An estimate (s_α^2) of the population variance can also be found from the between treatments sum of squares. However, this estimate is unbiased only if the null hypothesis is true, i.e. only if the population treatment means are all equal. Differences among the population means inflate this estimate and as shown in the next section a comparison of s_α^2 with s_e^2 by means of an F test is equivalent to testing the hypothesis of equality of means (μ_i).

While the algebra required to show that s_α^2 is an unbiased estimate is somewhat tedious in the case where the numbers of observations (n_i) per treatment are not equal, the theory presented so far in this text enables a proof to be given where the numbers are all equal.

Suppose then that

$$n_1 = n_2 = n_3 = \ldots = n_t \, (=n).$$

Then if

$$\mu_1 = \mu_2 = \mu_3 = \ldots = \mu_t \, (=\mu)$$

each treatment mean, $\bar{y}_{i.}$, is distributed with mean μ and variance σ^2/n. Therefore the treatment means are a random sample of size t from a population whose mean is μ and whose variance is σ^2/n. The members of this sample are $\bar{y}_{1.}, \bar{y}_{2.}, \ldots, \bar{y}_{t.}$ and may be used to find an unbiased estimate of the variance of the population of which they are members.

Thus an estimate of this variance (σ^2/n) is given by

$$\sum (\bar{y}_{i.} - \bar{y}_{..})^2/(t-1).$$

This is equivalent to saying that an estimate of σ^2 is

$$\sum (X_i - \bar{x}_.)^2/(n-1).$$

If $\sum (\bar{y}_{i.} - \bar{y}_{..})^2/(t-1)$ is an estimate of σ^2/n, then an estimate (s_α^2) of σ^2 is

$$\sum n(\bar{y}_{i.} - \bar{y}_{..})^2/(t-1).$$

Even without the restriction that the n_i are equal, it can be shown that an estimate of σ^2 is

$$s_\alpha^2 = \sum_{i=1}^{t} n_i (\bar{y}_{i.} - \bar{y}_{..})^2/(t-1)$$

$$= \text{Treatment mean square.}$$

The estimate s_α^2 is based on $(t-1)$ degrees of freedom since it has been calculated from a sample of size t and is unbiased only if the treatment means are all equal. If the means are not equal the expected value of s_α^2 is

$$\sigma^2 + n \sum \alpha^2/(t-1) = \sigma_\alpha^2$$

when all the n_i are equal to n and is

$$\sigma^2 + n_0 \sum \alpha_i^2/(t-1) = \sigma_\alpha^2$$

where $n_0 = \sum n_i - (\sum n_i^2/\sum n_i)$ when the n_i are not all equal. $n \sum \alpha_i^2/(t-1)$ or $n_0 \sum \alpha_i^2/(t-1)$ is the inflation which was referred to in the first paragraph of this section and which results from differences among the population means.

9.5   The Analysis of Variance Table (One-way)

The calculations for an analysis of variance are set out in Tables 9.2.

TABLE 9.2   *Analysis of variance (one-way)*

Source of variation	d.f.	Sum of squares	Mean square	F
Between treatments	$t-1$	$\sum n_i(\bar{y}_{i\cdot} - \bar{y}_{\cdot\cdot})^2 = B$	$B/(t-1) = s_\alpha^2$	s_α^2/s_e^2
Within treatments	$N-t$	$\sum\sum (Y_{ij} - \bar{y}_{i\cdot})^2 = W$	$W/(N-t) = s_e^2$	
Total	$N-1$	$\sum\sum (Y_{ij} - \bar{y}_{\cdot\cdot})^2$		

The entries in the sum of squares column are usually not found directly from the formulae presented in the above table but from formulae (9.2), (9.3) and (9.4).

From the previous section, the expected value of s_α^2 is a minimum and equals σ^2 when the treatment population means are equal. Differences in the μ_i tend to increase the between treatments mean square, and such differences are indicated by s_α^2 being greater than s_e^2. Hence if $s_\alpha^2 \leqslant s_e^2$, it is immediately concluded that there is no evidence upon which to reject the null hypothesis of equality of treatment means.

If, however, $s_\alpha^2 > s_e^2$, the F test is used to determine whether s_α^2 is significantly greater than s_e^2 and hence, by implication, the treatment population means are not all equal. This is a one-tailed test of significance, and the generally used levels of significance are 5% and 1%. The significance of the observed F's is frequently indicated by a single (5%) or double (1%) asterisk.*

The null and alternative hypotheses may be stated as

$$H_0: \sigma_\alpha^2 = \sigma^2; \qquad H_1: \sigma_\alpha^2 > \sigma^2$$

and so the probabilities as stated in the F table are used. If it is concluded that $\sigma_\alpha^2 > \sigma^2$, the original null hypothesis concerning the treatment means is rejected since the null and alternative hypo-

---

* Some experimenters even extend this convention and use three asterisks to indicate significance at the 0.1% or 0.001 level. The convention which was first referred to in section 6.6 is not confined to z or F tests but is widely used with all test statistics (z, t, F, χ^2 to quote only a few) and even with the statistic itself (e.g. $\bar{d} = 4.56*$, $b = 1.788**$, $r = 0.57*$).

theses concerning σ_α^2 and σ^2 are equivalent to

$$H_0 : \mu_1 = \mu_2 = \mu_3 = \ldots = \mu_t,$$
$$H_1 : \text{at least one mean different.}$$

In the table above, the terms 'between treatments' and 'within treatments' have been used. This was done as a natural sequence to the theory which was developed in the previous section. However for the animal nutritionist with the 60 cows divided into 4 different breeds, these terms would be replaced by 'between breeds' and 'within breeds', the degrees of freedom being respectively 3 and 56. Again, a plant breeder with different groups of inbred families would divide his total variation into 'between families' and 'within families'.

Example 9.1 To test the yielding ability of seven different varieties of wheat, 35 plots of approximately equal fertility were sown, 5 plots to each variety. The table below gives the yields of grain in bushels/acre. Assuming the data fit the mathematical model

$$Y_{ij} = \mu_i + \varepsilon_{ij}, \quad (i = 1, 2, \ldots, 7; j = 1, \ldots, 5)$$

where μ_i is the population mean for the ith variety and ε_{ij} are random error terms normally distributed with zero mean and variance σ^2, investigate whether the varietal population means are equal.

Variety	Yield (bushels/acre)					Varietal total ($Y_{i.}$)
A	10	10	15	14	17	66
B	9	10	13	15	14	61
C	11	12	10	13	16	62
D	10	11	12	14	13	60
E	12	11	13	16	15	67
F	15	12	14	17	19	77
G	11	12	13	15	18	69

Here

$$N = \sum n_i = 35.$$

[If a calculating machine is available, the data are used as they stand. If no machine is available, the working is simplified by a change of origin. Subtract, say 10, from each measurement and work with a new variable.

$$U_{ij} = Y_{ij} - 10.$$

The working for the variable U_{ij} is left as an exercise for the reader.]

Grand total ($Y_{..}$)	$= \sum\sum Y_{ij}$	$= 10+10+15+...+13+15+18 =$	462·0
General mean ($\bar{y}_{..}$)	$= Y_{..}/N$	$= 462 \div 35$	13·2
Correction term	$= Y_{..}^2/N$	$= 462^2 \div 35$	6098·4

Total S.S. $= \sum\sum Y_{ij}^2 = 10^2+10^2+... \quad + 15^2 + 18^2 =$ 6314
$\quad\quad - Y_{..}^2/N$ $-6098\cdot4$
$\qquad\qquad\qquad\qquad\qquad\qquad\qquad\qquad\qquad\qquad\qquad\qquad$ 215·6

Between varieties S.S. $= \sum Y_{i.}^2/n_i = (66^2+61^2+...+69^2)/5 \quad =$ 6140
$\quad\quad\quad\quad\quad - Y_{..}^2/N$ $-6098\cdot4$
$\qquad\qquad\qquad\qquad\qquad\qquad\qquad\qquad\qquad\qquad\qquad\qquad$ 41·6

Within varieties S.S. = Total S.S. − Between varieties S.S. $=$ 174·0

While the within varieties sum of squares has been obtained by subtracting the between varieties sum of squares from the total sum of squares, an overall check on the arithmetic could be made by calculating the within varieties sum of squares directly as

$$(10^2 + 10^2 + ... + 17^2 - 66^2/5) + (9^2 + 10^2 + ...14^2 - 61^2/5) + ... + (11^2 + 12^2 + ...$$
$$+ 18^2 - 69^2/5).$$

The check could be completed by noting that

Total S.S. = Between S.S. + Within S.S.

The analysis of variance is as shown.

Analysis of variance: yield (bushels/acre)

Source of variation	d.f.	Sum of squares	Mean square	F
Between varieties	6	41·6	6·93	1·12 n.s.
Within varieties	28	174·0	6·21	
Total	34	215·6		

The null and alternative hypotheses being studied are

$$H_0: \mu_1 = \mu_2 = \mu_3 = \mu_4 = \mu_5 = \mu_6 = \mu_7,$$

H_1: at least one pair of means is not equal.

Suppose the level of significance chosen is 0·05. From the analysis of variance, if the null hypothesis is true, the variance ratio follows an F distribution with degrees of freedom $v_1 = 6$ and $v_2 = 28$.

The observed value of $F(=1.12)$ has probability greater than 5% and is not significant. The null hypothesis of equality of varietal means is not rejected since the two estimates of variance are not sufficiently different to lead to rejection of the hypothesis that $\sigma_\alpha^2 = \sigma^2$.

Exercises

9.1   The following are the plot yields (Y_{ij}) in bushels/acre obtained from a wheat variety trial. The plots for each variety were randomly allocated throughout the field.

Plot number	1	2	3	4	5	6	7	8	9	10
Variety	C	A	E	B	A	D	E	C	E	C
Yield	23	18	17	21	17	24	18	23	18	25

Plot number	11	12	13	14	15	16	17	18	19	20
Variety	C	D	E	A	E	B	B	D	E	C
Yield	25	27	19	15	20	19	20	29	20	25

Plot number	21	22	23	24	25	26	27	28	29
Variety	A	C	E	B	C	A	D	B	D
Yield	19	26	21	23	28	21	31	17	29

(a) For this set of data, prepare a table similar to Table 9.1.

(b) From each Y_{ij} subtract $\bar{y}_{..}$ and then, using the terms $Y_{ij} - \bar{y}_{..}$, calculate $\sum_i \sum_j (Y_{ij} - \bar{y}_{..})^2$.

(c) From each varietal mean, $\bar{y}_{i.}$, subtract the general mean, $\bar{y}_{..}$ and, using the terms $\bar{y}_{i.} - \bar{y}_{..}$, calculate $\sum_i n_i (\bar{y}_{i.} - \bar{y}_{..})^2$.

(d) From each Y_{ij} subtract the varietal mean, $\bar{y}_{i.}$, and, using the terms $Y_{ij} - \bar{y}_{i.}$, calculate $\sum_i \sum_j (Y_{ij} - \bar{y}_{i.})^2$.

(e) Check that total S.S. = between S.S. + within S.S.

(f) Use formulae (9.2) and (9.3) to find the total S.S. and the between S.S. and observe that these sums agree respectively with those found in (b) and (c).

(g) Assuming that wheat yields are normally distributed with equal variance, carry out an analysis of variance to test the hypothesis of equality of varietal means.

9.2   The weight gains in arbitrary units, in a given time for four groups of lambs fed on four different diets are given in the following table. Initially five lambs were randomly allocated to each diet group. However as a result of sickness (not associated with diet) one lamb had to be removed from the second diet group and two from the third. Test the hypothesis that the mean weight gains for the four different diets are equal.

Diet 1	13	11	12	14	15
Diet 2	14	12	13	13	
Diet 3	16	15	14		
Diet 4	10	12	10	12	11

If the variable (weight gain) is transformed by the subtraction of 10, this analysis may be carried out without the use of a calculating machine. All sums of squares, mean squares and the observed F are unaltered by such a transformation.

9.3   For the data of Exercise 7.8 (p. 133), use the one-way analysis of variance to test the hypothesis that $\mu_1 = \mu_2$. Compare the observed F with the observed t as computed in Chapter 7 (note that $F_{1,\ 10} = t_{10}^2$).

9.6   Two Criteria of Classification: The Randomized Complete Block Design

In the examples and exercises just considered, the data were classified according to one factor only. However, the data which have to be analyzed in practice are often classified in several ways and it is only rarely that they are not classified in at least two ways.

Frequently in field and glasshouse trials, the treatments are arranged in blocks, each block containing every treatment. The treatments are randomized within each block, and the blocks are placed across any fertility gradient which exists. Such a design is called a randomized complete-block design. For a randomized complete-block design with five treatments and four blocks, the field lay-out might be as indicated.

Blocks	Treatments				
I	B	A	D	E	C
II	C	B	A	E	D
III	E	C	D	A	B
IV	A	E	B	D	C

Fertility gradient ↓

The randomized complete-block design is used not only in varietal or fertilizer trials but also finds considerable use in animal experiments where the blocks, for example, might be litters, weight or age classes. The term 'randomized complete-block' is used because the complete set of 'treatments' is randomized within each 'block'.

Let Y_{ij} be the yield of the ith treatment in the jth block. Then the yields for the general case of the two-way classification (n blocks, t treatments) may be represented as in Table 9.3.

TABLE 9.3   *Two-way classification of yields*

Treatment or variety	Block or class							Treatment	
	1	2	3	...	j	...	n	means	totals
1	Y_{11}	Y_{12}	Y_{13}	...	Y_{1j}	...	Y_{1n}	$\bar{y}_{1.}$	$Y_{1.}$
2	Y_{21}	Y_{22}	Y_{23}	...	Y_{2j}	...	Y_{2n}	$\bar{y}_{2.}$	$Y_{2.}$
3	Y_{31}	Y_{32}	Y_{33}	...	Y_{3j}	...	Y_{3n}	$\bar{y}_{3.}$	$Y_{3.}$
⋮	⋮	⋮	⋮		⋮		⋮	⋮	⋮
i	Y_{i1}	Y_{i2}	Y_{i3}	...	Y_{ij}	...	Y_{in}	$\bar{y}_{i.}$	$Y_{i.}$
⋮	⋮	⋮	⋮		⋮		⋮	⋮	⋮
t	Y_{t1}	Y_{t2}	Y_{t3}	...	Y_{tj}	...	Y_{tn}	$\bar{y}_{t.}$	$Y_{t.}$
Block means	$\bar{y}_{.1}$	$\bar{y}_{.2}$	$\bar{y}_{.3}$	...	$\bar{y}_{.j}$	...	$\bar{y}_{.n}$	$\bar{y}_{..}$ = general mean	
Block totals	$Y_{.1}$	$Y_{.2}$	$Y_{.3}$	...	$Y_{.j}$	...	$Y_{.n}$	$Y_{..}$ = grand total	

The essential difference between this arrangement and that of the one-way classification is that the variates in any one class or block have something in common in that they can be logically placed together and recognized as a definite unit. In each block there are the t treatments while for each treatment there are n plots, one in each block.

The mathematical model being studied is

$$Y_{ij} = \mu + \alpha_i + \beta_j + \varepsilon_{ij}$$

where μ is the overall population mean,

    α_i is the effect of the ith treatment (or variety),

    β_j is the effect of the jth block (or class),

    ε_{ij} are random errors which are normally and independently distributed with mean zero and variance σ^2.

The model may be written as

$$Y_{ij} = \mu_i + \beta_j + \varepsilon_{ij} \qquad \text{where } \mu_i = \mu + \alpha_i$$

and the following null hypothesis studied:

$$H_0 : \mu_1 = \mu_2 = \mu_3 = \ldots = \mu_t,$$

which is equivalent to

$$H_0 : \alpha_1 = \alpha_2 = \alpha_3 = \ldots = \alpha_t.$$

This is an extension of the method of paired comparisons and the paired t test where only two means. μ_1 and μ_2, were being considered. Here i takes the values $1, 2, \ldots, t$ and the β_j (the block effects) are analogous to the pair effects in the t test.

Here the general method of attack is the same as in the case of a single criterion of classification. By means of an F test, the variability in the treatment means is examined to see whether it is more than could be reasonably expected under the null hypothesis being studied.

With terminology similar to that used previously, it may be shown that

$$\sum_{i=1}^{t} \sum_{j=1}^{n} (Y_{ij} - \bar{y}_{..})^2 = \sum_{i=1}^{t} n(\bar{y}_{i.} - \bar{y}_{..})^2 + \sum_{j=1}^{n} t(\bar{y}_{.j} - \bar{y}_{..})^2$$

$$+ \sum_{i=1}^{t} \sum_{j=1}^{n} (Y_{ij} - \bar{y}_{i.} - \bar{y}_{.j} + \bar{y}_{..})^2 \qquad (9.5)$$

with corresponding degrees of freedom

$$nt - 1 = (t-1) + (n-1) + (t-1)(n-1).$$

This equation shows that
    Total variation = Variation between treatments
                    + Variation between blocks
                    + Residual or unexplained variation (error).

Compairing equation (9.5) with equation (9.1) shows that in both the completely randomized design and the randomized complete-block design, there is one term common in the components into which the total variation is partitioned. This is the between treatments variation. The within treatments variation in (9.1) has been partitioned into two parts in (9.5)—the between blocks and residual or error.

The residual in the randomized complete block design has fewer degrees of freedom than its counterpart in the completely random design. However, if the experiment is such that the 'block' term is expected to be large, this reduction in degrees of freedom is more than offset by the reduction in the error sum of squares. In practice the error mean square is usually found to be smaller than the within treatments mean square, and this results in the randomized complete block design being more efficient than the completely random design.

Returning to equation (9.5), the sums of squares in this equation, which are used in the analysis of variance, are calculated in the same way as those of the one-way classification:

$$\text{Total S.S.} = \sum_i \sum_j Y_{ij}^2 - Y_{..}^2/nt \quad (= A);$$

$$\text{Between treatments S.S.} = \sum_i Y_{i.}^2/n - Y_{..}^2/nt \quad (= B);$$

$$\text{Between blocks S.S.} = \sum_j Y_{.j}^2/t - Y_{..}^2/nt \quad (= C).$$

The error sum of squares is usually found by subtraction. It should be noted that the error sum of squares is the sum of terms of the type $(Y_{ij} - \bar{y}_{i.} - \bar{y}_{.j} + \bar{y}_{..})^2$ and as such can never be negative.

9.7 Estimates of the Population Variance (σ^2)

It can be shown that an unbiased estimate of the population variance is obtained (irrespective of the value of the treatment or block means) from

$$\frac{\sum\sum (Y_{ij} - \bar{y}_{i.} - \bar{y}_{.j} + \bar{y}_{..})^2}{(n-1)(t-1)} = \frac{\text{Error sum of squares}}{\text{Error degrees of freedom}}$$

$$= \text{Error mean square}$$

$$= s_e^2.$$

Also, if $\quad \alpha_1 = \alpha_2 = \ldots = \alpha_t$, i.e. if the treatment effects are the same,

$$\frac{\sum n(\bar{y}_{i.} - \bar{y}_{..})^2}{t-1} = \frac{\text{Treatment sum of squares}}{\text{Treatment degrees of freedom}}$$

$$= \text{Treatment mean square}$$

$$= s_\alpha^2$$

is an unbiased estimate of σ^2.

Thus the null hypothesis that the treatment means are equal may be tested by an F test with $(t-1)$ and $(n-1)(t-1)$ degrees of freedom, i.e.

$$F = \frac{s_\alpha^2}{s_e^2} = \frac{\text{Treatment mean square}}{\text{Error mean square}}.$$

It might be noted that under the hypothesis that the block or replicate effects are equal (i.e. $\beta_1 = \beta_2 = \ldots = \beta_n$), the block mean square is an unbiased estimate of σ^2. Thus an F test (block mean square divided by error mean square) could be used to examine the null hypothesis that the β's are equal. However, this test is rarely made because the experimenter is seldom interested in block effects *per se* and because the design is such that the experimenter would anticipate differences in the block means.

9.8 The Analysis of Variance Table (Two-way)

The analysis of variance (two-way) is set out in Table 9.4.

TABLE 9.4 *Analysis of variance (two-way)*

Source of variation	d.f.	Sum of squares	Mean square	F
Between blocks or replicates	$n-1$	C	$C/(n-1) = s_\beta^2$	s_β^2/s_e^2
Between treatments or varieties	$t-1$	B	$B/(t-1) = s_\alpha^2$	s_α^2/s_e^2
Error or residual	$(n-1)(t-1)$	$A-(B+C)$	$\{A-(B+C)\}/(n-1)(t-1)$ $= s_e^2$	
Total	$nt-1$	A		

Example 9.2 The data given in Example 9.1 are from an agricultural experiment set out in a randomized complete-block design. The experiment consisted of 5 blocks and 7 different varieties. Each block was divided into 7 plots and the plots of each block were assigned at random to the 7 varieties. The yields in bushels/acre are set out in Table 9.5.

TABLE 9.5 *Wheat yields in a randomized complete-block experiment*

Variety	Block I	II	III	IV	V	Varietal total ($Y_{i.}$)
A	10	10	15	14	17	66
B	9	10	13	15	14	61
C	11	12	10	13	16	62
D	10	11	12	14	13	60
E	12	11	13	16	15	67
F	15	12	14	17	19	77
G	11	12	13	15	18	69
Block total ($Y_{.j}$)	78	78	90	104	112	$462 = Y_{..}$

Discuss the significance of the variation of yield with each of the two criteria of classification.

The correction term, total sum of squares and between varieties sum of squares were found previously. The only additional term needed is

$$\text{Block S.S.} = \sum Y_{.j}^2/t - Y_{..}^2/nt$$

$$\doteq \frac{78^2 + 78^2 + \ldots + 112^2}{7} - \frac{462^2}{35}$$

$$= 134\cdot2.$$

The analysis of variance is:

Source of variation	d.f.	Sum of squares	Mean square	F
Blocks	4	134·2	33·55	20·2**
Varieties	6	41·6	6·93	4·2**
Error	24	39·8	1·66	
Total	34	215·6		

** = F value significant at the 1% level.

The tests of significance are as follows:

(a) $H_0 : \mu_1 = \mu_2 = \mu_3 = \ldots = \mu_7$,

H_1 : At least one pair of varietal means is not equal,

Level of significance $= 0.05$.

From the analysis of variance, if the null hypothesis is correct, the variance ratio (between varieties mean square/ error mean square) follows an F distribution with degrees of freedom $v_1 = 6$ and $v_2 = 24$. The value of F obtained has probability (p) less than 1%.

Hence $p < 0.05$, the chosen level of significance, and the null hypothesis is rejected.

It is concluded that the varietal means are not all equal.

(b) $H_0 : \beta_1 = \beta_2 = \ldots = \beta_5$,

H_1 : At least one pair of block means is not equal,

Level of significance $= 0.05$.

From the analysis of variance, if this null hypothesis is correct, the variance ratio (between blocks mean square/error mean square) follows an F distribution with degrees of freedom $v_1 = 4$ and $v_2 = 24$. The value of F obtained has probability (p) less than 1%.

Here $p < 0.05$, the chosen level of significance, and hence the null hypothesis is rejected.

It is concluded that the block means are not all equal.

Note: The differences among the block means increased the error mean square in the first example and resulted in a non-significant F test. The exclusion of this variability from the error sum of squares resulted in a significant F in the two-way analysis of variance. The design of the experiment, and hence the model which the data fit, determines the form of the analysis of variance.

9.9 Rejection of Null Hypothesis

In the previous example where a significant F was obtained, the hypothesis of equality of means was rejected. This is the usual procedure. However, the assumptions underlying the analysis of variance should not be overlooked. These are

(i) the linear additive model is the appropriate model—the treatment and environmental effects are additive;

(ii) the variances of yields in the different treatments and/or blocks are equal;

(iii) the populations sampled are normal.

Parts (ii) and (iii) relate to the experimental errors (ε_{ij}) which are assumed to be random, independently and normally distributed about zero mean and with a common variance.

Before using an F test to test the hypothesis of equality of treatment means, the experimenter should be reasonably certain that the data on which the analysis of variance is to be made can be regarded as satisfying these assumptions.

Eisenhart (1947), Cochran (1947) and Bartlett (1947) are three authors who have discussed these assumptions, the consequences when they do not hold, and the steps to be taken to arrange that the data to be analyzed satisfy the above requirements.

Tests are available to examine non-additivity. The most commonly used test is that given by Tukey (1949).

Heterogeneity of variance may be checked for by using Bartlett's (1937) test of homogeneity of variance. If there are large differences among the treatment or block means, these differences may indicate heterogeneity of error variances, since large means and large variances, small means and small variances often go hand in hand. The treatments with large means may well have large variances, while the treatments with small means might have small variances.

Where there is evidence of heterogeneity of variance and there is no theoretical reason for the mean and variance to be connected, the treatments have to be subdivided into groups having homogeneous variances and considered separately. Orthogonal coefficients are used to partition treatment and error sums of squares into orthogonal components in order to satisfy the assumptions underlying the analysis of variance. This topic is not considered here but is dealt with in most advanced texts on biometrical methods.

However, when dealing with Poisson and binomial variates, it should be remembered that there is a relationship between the mean and the variance. For the Poisson the mean is equal to the variance, and it might be anticipated that treatments having large means would have large variances and those having small means, small variances. For the binomial, variance $= n\pi(1-\pi)$ and again with differences in treatment means, differences in variance will occur. Thus in cases where the variate (e.g. small whole numbers, percentages, proportions) follows one or other of these distributions, there are theoretical reasons why the variate being studied must be transformed to a new variate which has relatively stable variance. It is of interest to note that these transformations also result in the normality assumption being more nearly satisfied. The transformations used most frequently are the square root, logarithmic and angular or inverse sine. Again the topic of transformations is not dealt with here. It is considered in some detail by Bartlett (1947) and Cochran (1947).

Finally, if the data satisfy the assumptions on which the analysis of variance is based, then both the treatment mean square and the error mean square are unbiased estimates of the population variance provided the null hypothesis of equality of treatment means holds. However, if the treatment means are not equal then the expectation of the treatment mean square is greater than the population variance. The treatment mean square becomes bigger as differences among the treatment population means become bigger. For this reason when a significant F is obtained, the hypothesis of equality of means is rejected.

9.10 Examination of Observed Treatment Means

When an F test rejects the null hypothesis

$$H_0: \mu_1 = \mu_2 = \ldots = \mu_t$$

it unfortunately gives no information about which of the means are causing the significance.

Suppose it is desired to examine more closely the difference between the observed means of any two varieties in the example studied, say D and G. The difference between the means of D and G is 1·8 bushels/acre. To test the significance of this a t test is used as follows:

$$t = \frac{\bar{y}_{7.} - \bar{y}_{4.}}{s\sqrt{2}/\sqrt{n}}$$

where s^2 is the unbiased estimate of the population variance obtained from the error mean square and n is the number of plots of each variety. This t has degrees of freedom equal to the error degrees of freedom.

The calculations are made as follows:

$$\text{s.d. per plot } (s) = \sqrt{(\text{Error mean square})}$$

$$= \sqrt{1 \cdot 66}$$

$$= 1 \cdot 29$$

$$\text{s.e. varietal mean} = \frac{s}{\sqrt{n}} = \frac{1 \cdot 29}{\sqrt{5}}$$

$$= 0 \cdot 577;$$

$$\begin{array}{l} \text{s.e. of difference} \\ \text{of 2 means} \end{array} = \frac{s \times \sqrt{2}}{\sqrt{n}} = 0 \cdot 577 \times \sqrt{2}$$

$$= 0 \cdot 816;$$

$$t = \frac{1 \cdot 8}{0 \cdot 816} = 2 \cdot 2.$$

The estimate of variance, s^2, is based on 24 d.f., for which the value of t is significant at the 5% level. It is concluded that the means of variety D and variety G are different.

In order to examine all the means in a systematic manner, it is common practice to order the means from greatest to least and to examine all possible differences among the set of means by a series of t tests. A quantity called the 'least significant difference' (l.s.d.) is calculated.

If the equation

$$t = \frac{\text{difference between two means}}{s\sqrt{2}/\sqrt{n}}$$

is rewritten in the form

$$\frac{ts\sqrt{2}}{\sqrt{n}} = \text{difference between two means,}$$

it can be seen that if t_1 is the value of t at the 5% level, then two means differing by more than $(t_1 s\sqrt{2})/\sqrt{n}$ differ significantly at the 5% level.

Thus, if it is desired to examine the hypotheses

$$H_0 : \mu_i = \mu_j \qquad (i = 1, 2, 3, \ldots, t; j \neq i)$$

at the 5% level, all observed differences are compared with

$$\text{l.s.d.} = \frac{s\sqrt{2t_1}}{\sqrt{n}}.$$

If the difference between any pair of means is greater than the l.s.d., the means are significantly different at the 5% level.

The calculations needed are usually set out below the analysis of variance in the following form:

$$\text{s.d. per plot} = s;$$
$$\text{s.e. varietal mean} = s/\sqrt{n};$$
$$\begin{array}{l}\text{s.e. of difference} \\ \text{of 2 varietal means}\end{array} = s\sqrt{2}/\sqrt{n};$$
$$\text{l.s.d. } (5\%) = \frac{st_1\sqrt{2}}{\sqrt{n}}.$$

For the example just considered, the steps are:

$$\text{s.d. per plot} = 1 \cdot 29;$$
$$\text{s.e. varietal mean} = 1 \cdot 29/\sqrt{5} = 0 \cdot 577;$$
$$\begin{array}{l}\text{s.e. difference of} \\ \text{2 varietal means}\end{array} = 0 \cdot 577/\sqrt{2} = 0 \cdot 816;$$
$$\text{l.s.d. } (5\%) = 0 \cdot 816 \times 2 \cdot 064$$
$$= 1 \cdot 68.$$

Then the differences among the means are set out in the table below and all differences compared in turn with 1·68.

| Variety | Mean | Differences from | | | | | |
		D	B	C	A	E	G
F	15·4	3·4	3·2	3·0	2·2	2·0	1·6
G	13·8	1·8	1·6	1·4	0·6	0·4	
E	13·4	1·4	1·2	1·0	0·2		
A	13·2	1·2	1·0	0·8			
C	12·4	0·4	0·2				
B	12·2	0·2					
D	12·0						

The difference F v D of 3·4 is greater than 1·68 and hence is significant; the difference F v B of 3·2 is greater than 1·68 and hence is significant. The only difference from F which is not significant if F v G. Next the differences from G are examined, then those from E. However the difference E v D is 1·4 and is less than 1·68. Hence it is non-significant and as all remaining differences are less than this, no further examination of the table of differences is necessary.

The results of the procedure may be summarized by underlining means which are not significantly different.

Variety	F	G	E	A	C	B	D
Mean	15·4	13·8	13·4	13·2	12·4	12·2	12·0

The differences between means of two varieties in a bracketed group are not significant at the 5% level. All pairs of means not appearing together in a common bracket differ significantly from one another.

9.11 Multiple Comparison Procedures and Error Rate Bases

In examining p treatment means there are $p(p-1)/2$ treatment differences. The bigger p becomes, the bigger the number of differences, and the more likely it is that the extreme differences will be found to be significant when judged by a t test. While it is understandable that an experimenter will frequently wish to examine interesting differences among the treatments suggested by the data themselves, it is important to realize the inherent dangers of this approach.

The indiscriminate use of the least significant difference procedure and the testing of differences suggested by the data (e.g. testing the largest mean against the smallest only because these two means are found to have the largest difference) has prompted many statisticians to consider what is known as the multiple comparisons problem. Different methods have been suggested for examining a set of observed means. Each of these has been proposed from a different base and this has led to some confusion and misuse of the various procedures.

The most frequently used of these procedures are the least significant difference (l.s.d.) procedure which was introduced in the previous section; Tukey's (1953) honestly significant difference (h.s.d.) procedure which uses the distribution of the range in a set of means; Scheffé's (1953) procedure; and Duncan's (1955) new multiple range procedure. These procedures are described in a number of texts, one of which is Federer (1955). Each has a different error rate base, and a decision to use one or another follows from an understanding of the appropriate error rate base.

While there are a number of error rate bases, three different bases determining an error rate (Federer (1961), Balaam and Federer (1965)) are now considered.

(i) Error rate per comparison

$$= \frac{\text{No. of erroneous inferences}}{\text{No. of inferences attempted}}.$$

This is the proportion of all comparisons expected to be erroneous when the null hypothesis is true. For example in using a 5% l.s.d. procedure, which has an error rate per comparison, if the null hypothesis were true and if 20 treatments were used in an experiment, 9 of the possible 190 ($= 20 \times 19/2$) differences would be expected to be significant. Balaam (1963) carried out a large simulation study which showed that when the number of means is small, the l.s.d. is a satisfactory procedure.

Fisher (1947) wrote:

'It is comparisons suggested subsequently, by a scrutiny of the results themselves, that are open to suspicion, for if the variants are numerous, a comparison of the highest with the lowest observed value, picked out from the results, will often appear significant, even from undifferentiated material. Properly, such unforeseen effects should be regarded only as suggestions for future experimentation, in which they can be deliberately tested.'

He went on to suggest a modification of the l.s.d. procedure in which the probability level chosen is $\alpha \Big/ \binom{p}{2}$ if p treatments are in the experiment. This modification has a different error rate base from the l.s.d. It is

(ii) Error rate per experiment

$$= \frac{\text{No. of erroneous inferences}}{\text{No. of experiments}}$$

N

which is the expected number of erroneous statements per experiment when the null hypothesis is true.

Tukey's h.s.d. procedure and Scheffé's procedure have a third type of error rate:

(iii) Experimentwise error rate

$$= \frac{\text{No. of experiments with 1 or more erroneous inferences}}{\text{No. of experiments}}$$

which is the proportion of experiments with one or more erroneous statements when the null hypothesis is true.

Federer (1961) lists the conditions that should hold before one or other procedure is used. He claims that for most biological experiments, the conceptual unit is the experiment, and hence he recommends the use of an experimentwise error rate. However, there are other statisticians who consider that frequently the comparison between two treatments is the conceptual unit, and in such cases they use an error rate per comparison and the l.s.d. procedure.

The purpose of this brief introduction to multiple comparisons is to indicate that there can be no general rule that an experimenter will always use a particular multiple comparisons procedure. The problem of sorting out differences among a set of observed means is a complex one. For one type of experimental situation one procedure might be used; for another type, a different procedure might be the correct one to use.

9.12   Presentation of Tables of Means and Standard Errors

The analysis of variance is one of the most widely used statistical techniques in biological research but it is to be hoped that the reader of this text does not have the impression that an analysis of variance is considered an adequate summary of the statistical analysis of data obtained in a biological experiment. It should be understood that the presentation of tables of means, standard errors and least significant differences is necessary. Finney (1955) writes:

'The analysis of variance far from being a more sophisticated presentation of the information, is usually no more than a scaffolding needed in preliminary study of the data: it does not require inclusion in the ultimate report, since its duty is accomplished when it has indicated what tables of means need to be discussed and has provided estimates of standard errors.'

Finney goes on to say that it would be unfortunate

'if novices in biometric practice were to imagine that abandonment of tables of means, the natural and common-sense summaries of experimental data, in favour of tables of analyses of variance was an advance in statistical sophistication at which they should aim.'

In the preface of this text are found the following words: 'Statistics is not biology; it is not the important research data or result, and is in no way the purpose of the study.' These words should be borne in mind when summaries of biometrical techniques used in a biological investigation are presented in reports of such study.

Exercises

9.4 In a randomized block experiment consisting of b blocks and t treatments, the bt yields can be arranged in a rectangular formation called a matrix.

$$
\begin{array}{l}
Y_{11} \; Y_{12} \ldots Y_{1j} \ldots Y_{1t} \\
Y_{21} \; Y_{22} \ldots Y_{2j} \ldots Y_{2t} \\
\vdots \\
Y_{i1} \; Y_{i2} \; \ldots Y_{ij} \ldots Y_{it} \\
\vdots \\
Y_{b1} \; Y_{b2} \; \ldots Y_{bj} \ldots Y_{bt}
\end{array}
$$

(a) Referring to the matrix, state in words the meaning of each of the following:

   (i) Y_{32};   (ii) Y_{i4};   (iii) $\displaystyle\sum_{j=1}^{t} Y_{3j}$;

   (iv) $\displaystyle\sum_{i=1}^{b} Y_{i2}$;   (v) $\displaystyle\sum_{j=1}^{t}\sum_{i=1}^{b} Y_{ij}$.

(b) Referring to the matrix of yields, write the symbolic expression for each of the following :

(i) The yield of the fourth treatment in the third block;

(ii) The yield of the third treatment in the ith block;

(iii) The yield of the jth treatment in the fourth block;

(iv) The sum of the yields of the first treatment in all blocks:

(v) The sum of the yields of all treatments in all blocks.

9.5   Prove that

$$\sum_{i=1}^{t} \sum_{j=1}^{b} (Y_{ij} - \bar{y}_{..})^2 = \sum_{i=1}^{t} b(\bar{y}_{i.} - \bar{y}_{..})^2 + \sum_{j=1}^{b} t(\bar{y}_{.j} - \bar{y}_{..})^2$$

$$+ \sum_{i=1}^{t} \sum_{j=1}^{b} (Y_{ij} - \bar{y}_{i.} - \bar{y}_{.j} + \bar{y}_{..})^2$$

where

$$\bar{y}_{i.} = \sum_{j=1}^{b} Y_{ij}/b \,;\; \bar{y}_{.j} = \sum_{i=1}^{t} Y_{ij}/t \,;\; \bar{y}_{..} = \sum_{i=1}^{t} \sum_{j=1}^{b} Y_{ij}/bt \,.$$

9.6   The analysis of variance of a randomized complete-block experiment with b blocks and t treatments is based on the equation presented in the previous exercise, where

$$Y_{ij} = \text{yield of } i\text{th treatment in } j\text{th block};$$
$$\bar{y}_{i.} = \text{mean yield of } i\text{th treatment};$$
$$\bar{y}_{.j} = \text{mean yield of } j\text{th block};$$
$$\bar{y}_{..} = \text{general mean}.$$

Write down a fictitious set of data with four treatments and two blocks such that the block sum of squares and treatment sum of squares are each different from zero but such that the error sum of squares is equal to zero.

9.7   In order to test the yielding ability of 5 different strains of wheat, 20 plots of approximately equal fertility were sown with the different strains. Each strain was planted in 4 plots, the distribution of the strains among the plots being random.

The yields of grain in bushels/acre are given below.

Strain A	27	29	26	24
Strain B	22	20	23	21
Strain C	25	28	27	23
Strain D	17	19	21	18
Strain E	23	26	20	21

Use the analysis of variance (one-way) to test whether the data show significant differences in the yields of the five strains.

By means of the t test, calculate the differences needed for significance between two strain means at both the 5 per cent and 1 per cent level.

Test the significance of the difference between the means of—
(a) strain A and strain E;
(b) strain C and strain D.

9.8   Fifteen pigs are divided into five groups according to weight. By a random process, each one of the three pigs in each group is allotted one of three different feeds. The weight gain in pounds by each pig for a fixed length of time is given in the table below.

Weight groups	Feed A	Feed B	Feed C
I	1·3	1·6	1·8
II	1·4	1·5	1·5
III	1·5	1·5	1·4
IV	1·3	1·4	1·6
V	1·5	1·7	1·5

Show that an F test gives a non-significant result (at the 5% level) for the test of the null hypothesis of equality of feed means.

9.9   Use the data in Example 7.4 (p. 138) and, by means of the analysis of variance, test the significance of the difference in yield shown by the two treatments.

Explain the equivalence of this method and the method of the t test used in Example 7.4.

9.10   Parker and Oakley (1963) compared the performance of eight bacterial strains with two serradella species, *O. sativus* and *O. compressus*, in the field. The results of their trial are given in the following table. The design used was a randomized complete-block. Use the analysis of variance technique for each species to test the hypotheses of equality of strain means.

Yields of green matter (g/plot). Species: O. sativus

Strain

Block	1	2	3	4	5	6	7	8
A	1942	2335	2834	1668	2090	2126	1900	1517
B	2266	1730	3128	1592	1989	2662	1944	2239
C	1715	1861	2344	2179	2646	2456	2676	2165
D	2202	1690	2350	1610	2370	2605	2325	2103

Model: $X_{ij} = \mu + \alpha_i + \beta_j + \varepsilon_{ij}$ $(i = 1,\ldots,8;\ \ j = 1,\ldots,4)$

$$\sum X_{ij} = \quad 69\,259, \qquad (\sum X_{ij})^2/32 = 149\,900\,283 \cdot 78,$$

$$\sum X_{ij}^2 = 154\,696\,799,$$

$$\sum X_{i.}^2 = 609\,598\,245, \qquad \sum X_{.j}^2 = 1\,200\,605\,033.$$

Yields of green matter (g/plot). Species: O. compressus

Strain

Block	1	2	3	4	5	6	7	8
A	1056	796	1301	789	1011	1306	894	1412
B	1010	1384	1524	1128	1164	1026	1450	908
C	1114	1080	990	775	1452	1272	1178	1196
D	1071	982	1138	800	775	454	1224	1416

Model: $X_{ij} = \mu + \alpha_i + \beta_j + \varepsilon_{ij}$ $(i = 1,\ldots,8;\ \ j = 1,\ldots,4)$

$$\sum X_{ij} = \quad 35\,076, \qquad (\sum X_{ij})^2/32 = 38\,447\,680 \cdot 50,$$

$$\sum X_{ij}^2 = \quad 40\,338\,554,$$

$$\sum X_{i.}^2 = 155\,485\,946, \qquad \sum X_{.j}^2 = 309\,212\,910.$$

For each species, if the F test is significant, examine the strain means using an l.s.d. test.

9.11   The set of data in Table 9.6 is part of the results of an experiment reported by Holford (1968). (See Exercise 7.15 for further details of the experiment.) Use the two-way analysis of variance to test the hypothesis of equality of the four treatments (P_0S_0, P_1S_0, P_0S_1, P_1S_1) means for the variables: (i) yield at first harvest; (ii) total yield after six harvests.

TABLE 9.6 *Experimental data for Exercise 9.11*

| Soil | Mean yields ($\bar{y}_{ij.}$) of 3 pots | | | | | | | |
| | Yield at first harvest | | | | Total yield: six harvests | | | |
	P_0S_0	P_1S_0	P_0S_1	P_1S_1	P_0S_0	P_1S_0	P_0S_1	P_1S_1
1	0·63	2·79	0·86	2·56	11·01	13·18	11·05	16·03
2	2·61	4·18	2·64	3·55	17·28	23·43	20·04	27·23
3	2·12	3·67	2·60	3·97	12·22	14·08	19·86	27·62
4	2·27	4·05	2·27	3·52	11·05	14·67	18·59	31·23
5	3·88	4·26	3·57	4·10	11·38	12·39	24·92	26·30
6	1·08	2·61	1·16	2·18	10·50	11·40	14·08	21·80
7	2·10	3·01	2·60	3·09	17·80	16·80	22·58	23·95
8	3·47	3·78	3·70	3·83	19·89	21·10	39·35	40·38
9	3·72	6·08	3·15	5·15	21·02	33·02	18·22	45·67
10	2·14	4·48	2·55	4·28	11·86	15·80	13·46	23·49

Assume the data satisfy the model

$$\bar{y}_{ij.} = \mu + \alpha_i + \tau_j + \varepsilon_{ij} \quad (i = 1, 2, \ldots, 10; \quad j = 1, 2, 3, 4)$$

where μ = general mean;

α_i = effect of the ith soil;

τ_j = effect of the jth treatment;

ε_{ij} are normally and independently distributed with zero means and variance σ^2.

Note: The yields (Y_{ijk}) of the individual pots from which the means $\bar{y}_{ij.}$ are computed would usually be assumed to satisfy the model

$$Y_{ijk} = \mu + \alpha_i + \tau_j + \gamma_{ij} + \varepsilon_{ijk}$$

where γ_{ij} are the effects of interaction between soils and treatments.

Factorial experiment models such as this and analyses associated with such models are not dealt with in this text.

REFERENCES

BALAAM, L. N. (1963). 'Multiple Comparisons – A Sampling Experiment'. *Aust. J. Statistics*, **5**, (2), 62–84.

BALAAM, L. N. and FEDERER, W. T. (1965) 'Error rate bases'. *Technometrics*, **7**, (2), 260–262.

BARTLETT, M. S. (1937). 'Some Examples of Statistical Methods of Research in Agriculture and Applied Biology', *J. Roy. Stat. Soc. Suppl.*, **4**, 137–183.

BARTLETT. M. S. (1947). 'The Use of Transformations', *Biometrics*, **3**, 39–52.

COCHRAN, W. G. (1947). 'Some Consequences When the Assumptions for the Analysis of Variance are not Satisfied', *Biometrics*, **3**, 22–38.

DUNCAN, D. B. (1955). 'Multiple Range and Multiple *F* Tests', *Biometrics*, **11**, 1–42.

EISENHART, C. (1947). 'The Assumptions Underlying the Analysis of Variance', *Biometrics*, **3**, 1–21.

FEDERER, W. T. (1955). *Experimental Design: Theory and Application.* Macmillan, New York.

FEDERER, W. T. (1961). 'Experimental Error Rates', *Proc. Amer. Soc. Hort. Sci.*, **78**, 605–615.

FINNEY, D. J. (1955). 'Letter to the Editor', *Biometrics*, **11**, 502–503.

FISHER, R. A. (1947). *The Design of Experiments.* Oliver & Boyd, Edinburgh.

FISHER, R. A. (1948). *Statistical Methods for Research Workers* (11th edition). Oliver & Boyd, Edinburgh; Hafner, New York.

HOLFORD, I. C. R. (1968). 'Nitrogen Fixation by Lucerne on Several Soils in Pot Culture', *Aust. J. Exper. Agric. & Animal Husb.*, **8**, 555–560.

PARKER, C. A. and OAKLEY, A. E. (1963). 'Nodule Bacteria for Two Species of Serradella—*Ornithopus sativus* and *Ornithopus compressus*', *Aust. J. Exper. Agric. & Animal Husb.*, **3**. (8). 9–10.

SCHEFFÉ, H. (1953). 'A Method for Judging all Contrasts in the Analysis of Variance', *Biometrika*, **40**, 87–104.

TUKEY, J. W. (1949). 'One Degree of Freedom for Non-additivity'. *Biometrics*. **5**, 232–242.

TUKEY, J. W. (1953). 'The Problem of Multiple Comparisons', Ditto, Princeton University.

COLLATERAL READING

FEDERER, W. T. (1955). *Experimental Design: Theory and Application.* Macmillan, New York. Chapters 2, 4 & 5.

GOULDEN, C. H. (1952). *Methods of Statistical Analysis.* John Wiley, New York. Chapter 5.

PARADINE, C. G. and RIVETT, B. H. P. (1969). *Statistical Methods for Technologists* (2nd edition). English Universities Press, London. Chapter 12.

STEEL, R. G. D. and TORRIE, J. H. (1960). *Principles and Procedures of Statistics.* McGraw-Hill, New York. Chapters 7 & 8.

SNEDECOR, G. W. and COCHRAN, W. G. (1967). *Statistical Methods* (6th edition). Iowa State University Press, Ames, Iowa. Chapters 10 & 11.

WEATHERBURN, C. E. (1947). *Mathematical Statistics.* Cambridge University Press, Cambridge. Chapter 11.

CHAPTER 10

Simple Linear Regression and Correlation

10.1 Introduction

In Chapters 1 to 9 the variation in a single variable was studied. However, in biological and agricultural research the problem frequently arises of relating variation in one variable with the variation occurring in one or more other variables. For example it may be desired to study the way in which wheat yields vary with change in plant density, or the manner in which animal weights vary with protein level of feed, or how the number of eggs laid by a given species of mosquito varies with the saline content of the water in which it is living.

In many such problems, it may be possible to distinguish between the variable of primary interest and one or more secondary variables which are of interest only in so far as they affect the main one. In dealing with these it is conventional to refer to the main one as the *dependent* variable and to denote it by Y, and to refer to the secondary variables as the *independent* variables and to denote them by $X_1, X_2, \ldots, X_k$, or if there is only one independent variable, by X. Thus, in studying the way in which yields of wheat vary in relation to change of amount of fertilizer applied yields would constitute the dependent variable Y, and the fertilizer levels the independent variable X.

In general, regression and correlation problems can be divided into two categories—*simple* and *multiple*, depending on whether problems concerning two (simple) or more than two (multiple) variables are being studied. Simple regression problems may in turn be classified into linear and non-linear types according to whether or not a mathematical function representing a straight line can be used to describe the relationship between the two variables. In other words, a linear regression problem is one in which the same change in the dependent variable Y can be expected for a given change in the independent variable X, irrespective of the value of X.

In this text only simple linear regression and correlation problems will be studied.

Example 10.1 As an illustration of a simple linear regression problem, consider the regression of the following weights of 7 calves (expressed in tens of pounds) on their ages (expressed in months).

Weight (Y)	Age (X)
4	0
7	2
14	3
10	4
20	6
28	8
36	12
$\sum Y = 119$	$\sum X = 35$
$\bar{y}. = 17$	$\bar{x}. = 5$

As a starting point in this regression problem, the relationship between the dependent variable (weight) and the independent variable (age) may be illustrated in the form of a *scatter diagram* in which the weights are plotted against the ages (Figure 10.1).

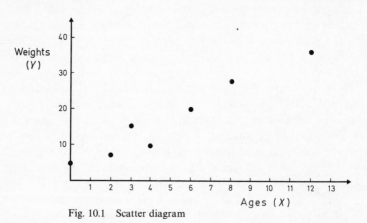

Fig. 10.1 Scatter diagram

It is evident from the scatter diagram that there is a linear relationship between weight and age, that weight is affected by age or, speaking in statistical terms, that there is a *linear regression of weight on age.*

10.2 Simple Linear Regression

Consider the problem of finding the straight line which is the best fit for a series of n points (X_i, Y_i). The term 'best fit' is interpreted in accordance with the principle of least squares. The constants a and b of the equation

$$Y = a + bX \tag{10.1}$$

are chosen so as to minimize the sum of squares of the deviations of the Y_i (the actual values of Y for each X_i) from the values ($\hat{Y}_i$) estimated by the equation (10.1),

$$\hat{Y}_i = a + bX_i.$$

The sum of squares of deviations, D, is given by

$$D = \sum_{i=1}^{n} (Y_i - \hat{Y}_i)^2$$

$$= \sum_{i=1}^{n} (Y_i - a - bX_i)^2.$$

Expanding the right-hand side of this equation gives

$$D = S_{yy} + n(a - \bar{y}_. + b\bar{x}_.)^2 + S_{xx}\left[b - \frac{S_{yx}}{S_{xx}}\right]^2 - \frac{S_{yx}^2}{S_{xx}} \tag{10.2}$$

where $S_{yy} = \sum (Y - \bar{y}_.)^2$; $S_{yx} = \sum (Y - \bar{y})(X - \bar{x})$; $S_{xx} = \sum (X - \bar{x}_.)^2$.

Two terms on the right-hand side of this equation (the second and third) are functions of a and b. The values of a and b which minimize D are obtained by setting these two terms equal to zero and hence

$$a - \bar{y}_. + b\bar{x}_. = 0,$$
$$b - S_{yx}/S_{xx} = 0.$$

Rewriting these two equations gives

$$\bar{y}_. = a + b\bar{x}_., \tag{10.3}$$
$$b = S_{yx}/S_{xx}. \tag{10.4}$$

The line of best fit is known as the *regression line*. Equation (10.3) shows that the regression line passes through the mean $(\bar{x}_., \bar{y}_.)$ and equation (10.4) gives the slope of the regression line. The coefficient b is called the *regression coefficient*. While this coefficient is defined by the equation (10.4), it is calculated from the following formula:

$$b = \frac{\sum YX - (\sum Y \sum X)/n}{\sum X^2 - (\sum X)^2/n}.$$

10.3 The Distribution of b: Test of Significance of Regression Coefficient

Just as the one way analysis of variance model was an extension of the linear additive model presented in section 6.3, the regression model may be thought of as an extension of the model for the one-way analysis of variance. In the one-way analysis of variance a series of Y populations with means μ_i are considered. The differences among the means may be examined and the hypothesis of equality of the μ_i tested by the analysis of variance. If the treatments are levels (X) of a quantitative factor it may be more reasonable to assume that the population means are dependent on X and to study the form of the dependence relationship.

The dependence may be indicated by rewriting the one-way analysis of variance model

$$Y_{ij} = \mu + \alpha_i + \varepsilon_{ij} \quad (i = 1, 2, \ldots, t; j = 1, 2, \ldots, r)$$

as

$$Y_{ij} = \mu_{y.x} + \varepsilon_{ij}$$

where $\mu_{y.x}$ indicates that the population means are dependent on the level of X.

In regression problems it is frequently the case that there is only one observation of Y for each value of X. Then if there are n observations altogether the model is written as

$$Y_i = \mu_{y.x} + \varepsilon_i \quad (i = 1, 2, \ldots, n).$$

Here there are n population means being investigated and, if these are assumed to be linearly related to X so that

$$\mu_{y.x} = \alpha + \beta X,$$

the regression model becomes

$$Y_i = \alpha + \beta X_i + \varepsilon_i$$

where α and β are parameters and ε_i are independently and normally distributed with mean zero and variance σ^2.

The parameters α and β are estimated as in the previous section. $\hat{\alpha}$ (or the intercept a) is the estimate of the Y population mean when $X_i = 0$ and since the *regression line* is

$$Y_i - \bar{y}_. = \hat{\beta}(X_i - \bar{x}_.)$$

the estimate of the population mean for the array of Y when $X_i = \bar{x}_.$ is $\bar{y}_.$.

Figure 10.2 illustrates the series of population means which are studied in a typical simple linear regression problem.

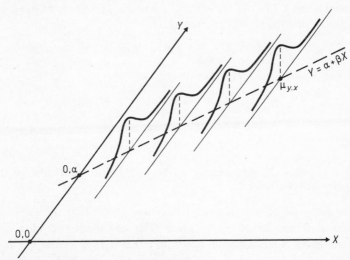

Fig. 10.2   An illustration of the regression model

The sample regression coefficient which estimates the population regression coefficient (β) is

$$b = \frac{\sum (Y_i - \bar{y}_.)(X_i - \bar{x}_.)}{\sum (X_i - \bar{x}_.)^2}$$

$$= \frac{\sum Y_i(X_i - \bar{x}_.)}{\sum (X_i - \bar{x}_.)^2}$$

$$= \frac{Y_1(X_1 - \bar{x}.)}{S_{xx}} + \frac{Y_2(X_2 - \bar{x}.)}{S_{xx}} + \ldots + \frac{Y_n(X_n - \bar{x}.)}{S_{xx}}.$$

By considering b as a linear function of this type and using Theorems 3.1, 3.2 and 3.4, it can be shown that b is distributed normally with mean β and variance σ^2/S_{xx}.

These results may be used to test null hypotheses concerning the value of β, e.g. $\beta = \beta_0$. If σ^2 were known a z test could be used. However, in practice σ^2 is unknown and so the t distribution is used.

The following t test is made

$$t = \frac{b - \beta_0}{s/\sqrt{\sum (X_i - \bar{x}.)^2}}$$

where $s^2 = \sum (Y - \hat{Y})^2/(n-2)$ and t has $(n-2)$ degrees of freedom.

The estimated standard error of b is

$$s_b = s/\sqrt{\sum (X_i - \bar{x}.)^2}.$$

The $(1 - \alpha)$ confidence interval for β is

$$\{b - t_{n-2, \frac{1}{2}\alpha} s_b\} \leqslant \beta \leqslant \{b + t_{n-2, \frac{1}{2}\alpha} s_b\}.$$

10.4   Testing the Significance of the Regression Coefficient by the Analysis of Variance

The estimate of variance s^2, which was defined in the previous section, is not found by squaring each deviation, summing and dividing by $(n-2)$. Instead, equation (10.2) is used to find the sum of squares of deviations from the regression line. Since the regression line is determined so that the second and third terms on the right-hand side of equation (10.2) are zero, this equation becomes

$$D = S_{yy} - S_{yx}^2/S_{xx} \tag{10.5}$$

Deviations S.S. = Total S.S. − Regression S.S.

The deviations sum of squares is found by subtracting the regression sum of squares from the total sum of squares. The total sum of squares is found in the usual way, $\sum Y^2 - (\sum Y)^2/n$ while the regression sum of squares is obtained by multiplying the regression coefficient by the sum of products, i.e.

$$b \sum (Y - \bar{y}_.)(X - \bar{x}_.) = b(\sum YX - \sum Y \sum X/n).$$

The deviations sum of squares divided by $(n-2)$ is an unbiased estimator of σ^2 irrespective of the value of β. If β is zero, i.e. if there is no regression of Y on X, the regression sum of squares also has an expected value equal to σ^2. However, if β is different from zero the regression sum of squares is inflated. This provides the basis for the following analysis of variance to test $\beta = 0$.

Analysis of variance

Source of variation	d.f.	Sum of squares	Mean squares	F
Regression function or reduction due to regression	1	bS_{yx}	$bS_{yx} = s_1^2$	s_1^2/s^2
Deviations from regression function	$n-2$	D	$D/(n-2) = s^2$	
Total	$n-1$	$\sum (Y - \bar{y}_.)^2$		

If the mean square due to the regression function is significantly greater than that due to deviations from this function, the null hypothesis that $\beta = 0$ is rejected and it is concluded that there is a linear relationship between the variables of the type indicated by the regression equation.

While the analysis of variance may be used to test only the hypothesis that $\beta = 0$ and is not as general as the test given in the previous section, it presents a compact method of calculating s^2. Further, it will be found to be useful when testing the significance of the correlation coefficient (see section 10.7).

It might be noted that if $\beta_0 = 0$, the test in the previous section becomes

$$t = \frac{b\sqrt{\sum (X_i - \bar{x}_.)^2}}{s}.$$

Then

$$t^2 = \frac{b^2 \sum (X_i - \bar{x}_.)^2}{s^2}$$

$$= \frac{b \sum (Y_i - \bar{y}_.)(X_i - \bar{x}_.)}{s^2}$$

$$= F_{1, n-2}$$

which is the F test used in the analysis of variance.

Example 10.2 Find the line of regression of weight on age for the data on weights and ages of calves given in Example 10.1. Use the analysis of variance to test the hypothesis that $\beta = 0$ In previous research of this type it had been found that the regression of weight on age was 25 lb/month. Use the t test to determine whether the observed regression coefficient differs significantly from 25 lb/month and calculate the 95% confidence interval for the regression coefficient.

To obtain the regression coefficient it is necessary to calculate the sum of products, $\sum (Y - \bar{y}_.)(X - \bar{x}_.)$, and the sum of squares, $\sum (X - \bar{x}_.)^2$. In addition as an analysis of variance is required, the sum of squares, $\sum (Y - \bar{y}_.)^2$ has to be found. From Example 10.1, $n = 7, \sum Y = 119, \bar{y}_. = 17, \sum X = 35, \bar{x}_. = 5$. Then

$$\sum Y^2 = 2841 \qquad \sum YX = 872 \qquad \sum X^2 = 273$$

$$(\sum Y)^2/n = 2023 \qquad (\sum Y)(\sum X)/n = 595 \qquad (\sum X)^2/n = 175$$

$$\sum (Y - \bar{y}_.)^2 = 818 \qquad \sum (Y - \bar{y}_.)(X - \bar{x}_.) = 277 \qquad \sum (X - \bar{x}_.)^2 = 98$$

$$b = \frac{\sum (Y - \bar{y}_.)(X - \bar{x}_.)}{\sum (X - \bar{x}_.)^2} = \frac{\sum YX - (\sum Y)(\sum X)/n}{\sum X^2 - (\sum X)^2/n}$$

$$= 277/98 = 2\cdot827 \text{ (tens of pounds/month)}$$

The regression equation is

$$Y - \bar{y}_. = b(X - \bar{x}_.),$$
$$Y - 17 = 2\cdot827(X - 5);$$
$$Y = 2\cdot827X + 2\cdot865.$$

Reduction due to regression $= bS_{yx}$

$$= (S_{yx})^2/S_{xx}$$

$$= 277^2/98 = 782.95.$$

The following is the analysis of variance which is used to test the hypothesis that $\beta = 0$.

Source of variation	d.f.	Sum of squares	Mean square	F
Reduction due to regression	1	782.95	782.95	111.69**
Deviations or residual	5	35.05	7.01	
Total	6	818.00		

** $= F$ value significant at the 1% level.

The null hypothesis that $\beta = 0$ is rejected because the observed value of F is significant at the 1% level. It is concluded that there is a linear relationship between the variables. The best estimate of this relationship is given by the regression equation.

The standard error of the regression coefficient is

$$s_b = \left[\frac{s^2}{\sum(X - \bar{x}.)^2} \right]^{\frac{1}{2}}$$

$$= \left[\frac{7.01}{98} \right]^{\frac{1}{2}}$$

$$= \sqrt{0.0715}$$

$$= 0.267 \text{ tens of pounds}$$

For $H_0: \beta = 25$ lb; $H_1: \beta \neq 25$ lb.

$$t = \frac{2.827 - 2.5}{0.267}$$

$$= \frac{0.327}{0.267}$$

$$= 1.22 .$$

o

This observed t which has 5 degrees of freedom is not significant. There is insufficient evidence to reject the hypothesis that $\beta = 25$ lb/month.

The 95% confidence interval for β is

$$\{b - t_{0.025,5}s_b\} \leqslant \beta \leqslant \{b + t_{0.025,5}s_b\},$$

$$\{2.827 - 2.571 \times 0.267\} \leqslant \beta \leqslant \{2.827 + 2.571 \times 0.267\},$$

$$2.141 \leqslant \beta \leqslant 3.513.$$

10.5 Scatter Diagrams

In Figures 10.3 (a) and (b) examples are given of scatter diagrams for samples from populations in which the variates are linearly related. In (a), the slope of the regression line is positive, while in (b), the slope is negative. However, in both diagrams the degree of association is approximately the same.

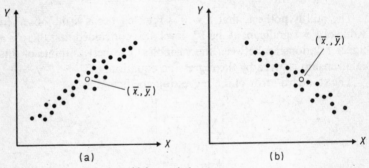

Fig. 10.3   Scatter diagrams—high association

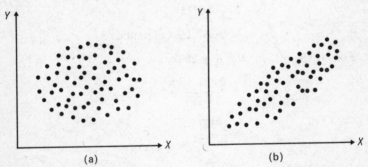

Fig. 10.4   Two more scatter diagrams

In contrast to such scatter diagrams, the points or dots may be distributed uniformly, as in Figure 10.4 (a), over the whole of the sample range and may show no tendency to cluster around a curve or to occupy a limited region. In this case the variates are said to be statistically independent. An intermediate stage between Figures 10.3 (a) and 10.4 (a) is that of 10.4 (b) in which the dots are not distributed at random all over the range of the sample but occupy a reasonably well defined region. In order to differentiate between situations as presented in Figures 10.3 (a) and (b), 10.4 (a) and (b), a measure of association between two variables is necessary. This is considered in the next section.

Exercises

10.1   (a), (b) and (c) are three sets of paired observations. For each set. find the linear regression of Y on X. Draw the three regression lines on three scatter diagrams. Do the slopes of the regression lines have the same sign? Test the significance of each of the three regression functions.

(a) Y	1	2	4	5	5	7
X	-3	-2	-1	1	2	3
(b) Y	-2	-1	0	1	2	
X	3	4	6	8	9	
(c) Y	7	5	4	0	2	0
X	-3	-1	1	3	5	7

10.2   In a fertilizer trial the yields obtained and the amounts of fertilizer applied are

Yield of barley (bushels/acre)	12	17	20	24
Amount of fertilizer (lb/acre)	90	180	270	360

Calculate the regression equation of yield on amount. Use the t test to test the hypothesis that $\beta = 4$ bushels per acre per hundred pounds of fertilizer.

10.3   A spectrophotometer is to be calibrated to measure the concentration of a certain drug (chloramphenicol) in blood serum. For this purpose five standard solutions are used, and their concentrations and the corresponding readings are shown in the following table:

Concentrations X (μg ml^{-1})	20	40	60	80	100
Readings Y		6	12	16	22	29

Draw a scatter diagram for the data.

Obtain the equation of the regression line of Y on X. (Work in units of U and V where $20U = X - 60$ and $V = Y - 17$).

Plot the regression line on the scatter diagram.

Test the significance of the regression by means of the analysis of variance.

10.6   The Simple Linear Correlation Coefficient

In Chapter 5, the univariate normal distribution and its density function were considered. The density function for the bivariate normal distribution is

$$\phi(X, Y) = \frac{1}{2\pi\sigma_X\sigma_Y(1 - \rho^2)^{\frac{1}{2}}} \exp T$$

where $T = -\dfrac{1}{2(1 - \rho^2)} \left\{ \left(\dfrac{X - \mu_X}{\sigma_X}\right)^2 - 2\rho\left(\dfrac{X - \mu_X}{\sigma_X}\right) \left(\dfrac{Y - \mu_Y}{\sigma_Y}\right)\right.$

$$\left. + \left(\frac{Y - \mu_Y}{\sigma_Y}\right)^2\right\}$$

and μ_X, μ_Y, σ_X, σ_Y, ρ are parameters; σ_X, $\sigma_Y > 0$, $-1 < \rho < +1$. The parameters μ_X, μ_Y are the means, and σ_X, σ_Y are the standard deviations, of the X and Y populations respectively; ρ is called the *correlation coefficient*.

In biological research, many types of paired measurements (X, Y) have bivariate normal distributions or distributions approximating this.

The bivariate normal surface, like the univariate normal, is bell shaped, and any cross-section parallel to the (X, Y) plane is either an ellipse or a circle. The centres of all such ellipses lie upon the vertical line through the mean point (μ_X, μ_Y) while the axes of these ellipses lie in two vertical planes perpendicular each to each. If ρ is near ± 1, and $\sigma_Y = \sigma_X$, the ellipses will be thin and the major axes will be much greater than the minor axes. As ρ gets closer to zero, the ellipses become closer to circles, again in those cases where the variance of X is equal to the variance of Y. The effect of differences

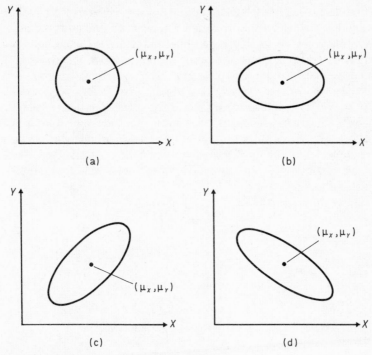

Fig. 10.5   Some bivariate normal contour ellipses
(a) $\rho = 0, \sigma_x = \sigma_y$ (b) $\rho = 0, \sigma_x > \sigma_y$
(c) $1 > \rho > 0, \sigma_x = \sigma_y$ (d) $-1 < \rho < 0, \sigma_x = \sigma_y$

in the values of the parameters $(\sigma_X, \sigma_Y, \rho)$ on the cross-sectional or contour ellipses is indicated in Figure 10.5.

The parameter ρ is a measure of the association between the two variables (X, Y). From a random sample (X_i, Y_i) of size n from a bivariate normal population an estimate of ρ is obtained by

$$r = \frac{\sum (Y_i - \bar{y})(X_i - \bar{x})}{\sqrt{[\sum (Y_i - \bar{y})^2 \sum (X_i - \bar{x})^2]}}$$

This sample correlation coefficient is given the same sign as $\sum (Y_i - \bar{y})(X_i - \bar{x})$ and hence has the same sign as the regression coefficient. It can be shown that $-1 \leqslant r \leqslant +1$. Further if $r = +1$ or -1, the sum of squares of deviations, D, from the regression line is zero. Each deviation is zero and all the points lie on the regression line. There is a linear functional relationship between the variables Y and X, giving perfect correlation. The nearer r^2 is to unity the closer are the points to the regression line, and hence the magnitude of r may be taken as a measure of the degree to which the association between the variables approach a linear functional relationship. When r is zero the variables are usually described as *uncorrelated*. It should be emphasized at this early stage that the correlation coefficient is a measure of association and not a measure of cause and effect. Many examples can be found of two variables for which r is quite large but for which there could be no cause and effect.

By analogy with the sample variance, the sample covariance is defined to be

$$\text{cov}\,(Y, X) = s_{yx} = \frac{1}{n-1} \sum\,(Y - \bar{y}_.)(X - \bar{x}_.)$$

$$= \frac{1}{n-1}\left[\sum YX - \frac{\sum Y \sum X}{n}\right].$$

Hence the correlation coefficient r is given by

$$r = \frac{\text{covariance}\,(Y, X)}{\sqrt{[\text{variance}(Y) \times \text{variance}(X)]}}.$$

10.7   Tests of Significance for the Correlation Coefficient

The analysis of variance may be used to test the hypothesis that the correlation coefficient is zero. To test the significance of a correlation coefficient is to examine whether the data of the sample indicate any degree of association of the variables of the type represented by the regression equation.

Returning to equation (10.5) it can be shown that

$$\text{Deviations S.S.} = (1 - r^2)S_{yy},$$
$$\text{Regression S.S.} = r^2 S_{yy} = bS_{yx}.$$

If $\rho = 0$, i.e. if the two variables are uncorrelated, both $r^2 S_{yy}$ and $(1 - r^2)S_{yy}/(n-2)$ are unbiased estimates of σ_y^2. However if $\rho \neq 0$, $r^2 S_{yy}$ has an expected value greater than σ_y^2 and is no longer an unbiased estimate of the variance. Thus an F test of the form $r^2 S_{yy}/\{(1 - r^2)S_{yy}/(n-2)\}$ can be made to test the hypothesis of no correlation. This test may be set out in an analysis of variance whose form is identical with that in section 10.4.

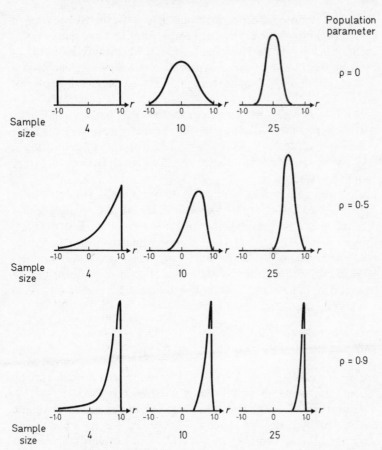

Fig. 10.6   Sampling distribution of the correlation coefficient (r) for three different population parameters (ρ), and three different sample sizes

The F test (for $\rho = 0$) in the analysis of variance can be written as a test. The t test is

$$t = \frac{r\sqrt{(n-2)}}{\sqrt{(1-r^2)}}. \tag{10.6}$$

This t variate has $n-2$ degrees of freedom.

The sampling distribution of r depends on the population correlation coefficient ρ and the sample size n (see Figure 10.6). The form of this distribution is important when making tests of significance and calculating confidence intervals. If $\rho = 0$, the distribution is symmetrical (irrespective of the sample size) and approaches normality when n is large. However the test of significance indicated by equation (10.6) can be made irrespective of sample size since the t in that equation is the square root of an F with degrees of freedom 1 and $(n-2)$.

It will be appreciated that equation (10.6) may be used in conjunction with the t table to determine for various sample sizes the value of r which would be significant at different levels of significance. A table giving significant correlation coefficients is to be found in Fisher and Yates (1963).

If ρ is not zero the distribution of r is not normal. However, from the results of Fisher (1921) and David (1938) tests of significance may be made and confidence intervals estimated. Fisher (1921) showed that $0.5 \ln\{(1+r)/(1-r)\}$ is approximately normally distributed with mean $0.5 \ln\{(1+\rho)/(1-\rho)\}$ and standard deviation $\sigma_z = 1/\sqrt{(n-3)}$ regardless of the value of ρ (ln is the logarithm to base e). Thus when making tests of significance for $\rho = \rho_0$ (not zero) it is common practice to calculate

$$Z_0 = 0.5 \ln\{(1+\rho_0)/(1-\rho_0)\}$$
and $$Z = 0.5 \ln\{1+r)/(1-r)\}$$

and to consider $(Z - Z_0)\sqrt{(n-3)}$ as a standardized normal variate, z. This approximation is sufficiently good for $n \geqslant 20$.

This transformation may also be used to set up approximate confidence intervals. The lower and upper limits in the transformed scale are $Z - z\sigma_z$ and $Z + z\sigma_z$, where z is a standardized normal variate and $\sigma_z = 1/\sqrt{(n-3)}$.

On retransforming these limits the confidence interval for the correlation coefficient is obtained. These will not be symmetrically centred around the sample estimate r. The limits are not of the form '$r \pm$ some quantity' as has been the case in all previous confidence intervals, e.g.

$$\bar{x} \pm z\sigma/\sqrt{n}, \quad \bar{x} \pm ts/\sqrt{n}, \quad b \pm t \text{ s.e. of } b.$$

Further, the Z transformation may be used to test the difference between two observed correlation coefficients. Thus if two independent random samples of size n_1 and n_2 have correlation coefficients of r_1 and r_2 and it is desired to test the hypothesis that $\rho_1 = \rho_2$,

Z_1 and Z_2 are calculated and $Z_1 - Z_2$ is treated as a normal variate with zero mean and standard deviation

$$\sqrt{\left(\frac{1}{n_1 - 3} + \frac{1}{n_2 - 3}\right)}.$$

If

$$(Z_1 - Z_2) \bigg/ \sqrt{\left(\frac{1}{n_1 - 3} + \frac{1}{n_2 - 3}\right)}$$

is greater than 1·96 or less than −1·96, the hypothesis is rejected at the 5% level; if it is greater than 2·58 or less than −2·58, the observed difference is significant at the 1% level.

It is important in applying the test outlined in the previous paragraph to remember that the two samples should be independent. For example if a random sample of animals is chosen and the correlation between two measurements (e.g. weight and girth) is obtained at one time and then at a later date the correlation between the two measurements is obtained for the same group of animals, this test should not be used since the two samples are not independent.

In order to complete the test of significance or to calculate the confidence intervals as outlined in the previous paragraphs, all that is necessary is a table converting r's to Z's. Table 10.1 is such a table.

TABLE 10.1   *Transformation of r to Z*

r	0·000	0·020	0·040	0·060	0·080
0·1	0·100	0·121	0·141	0·161	0·182
0·2	0·203	0·224	0·245	0·266	0·288
0·3	0·310	0·332	0·354	0·377	0·400
0·4	0·424	0·448	0·472	0·497	0·523
0·5	0·549	0·576	0·604	0·633	0·662
0·6	0·693	0·725	0·758	0·793	0·829
0·7	0·867	0·908	0·950	0·996	1·045
0·8	1·099	1·157	1·221	1·293	1·376
0·9	1·472	1·589	1·738	1·946	2·298

For small samples of less than 20, Fisher's transformation to the normal is not especially good. However, for various sample sizes less than 400, David (1938) obtained the distribution of r and produced charts from which may be estimated confidence limits for ρ when the sample size is small. The confidence interval may then be used to make a test of significance by noting whether the hypothesized value of ρ lies inside or outside the confidence interval.

Example 10.3 For the data given in Example 10.1, find the correlation coefficient and test the hypothesis that $\rho = 0$. Estimate the proportion of variation in weight of calves which is accounted for by the linear relationship of weight on age.

From Example 10.2,

$$\sum (Y - \bar{y}_.)^2 = 818,$$
$$\sum (Y - \bar{y}_.)(X - \bar{x}_.) = 277,$$
$$\sum (X - \bar{x}_.)^2 = 98,$$
$$r = \frac{277}{\sqrt{(818 \times 98)}} = 0\cdot98.$$

The analysis of variance given in Example 10.2 tests the hypothesis that $\rho = 0$. Since the observed F is highly significant, this hypothesis is rejected.

Since the regression sum of squares divided by the total sum of squares is r^2, the proportion of the total variation is $0\cdot98^2 = 0\cdot96$ or 96%. Used in this way, r^2 is known as the *coefficient of determination*.

Example 10.4 Cassie (1959) used replicate plankton samples in examining the correlations between the counts of four species— *Temora turbinata* (Dana), *Acartia clausii* Giesbrecht, *Oithona similis* Claus, *Coscinodiscus centralis* Ehrenberg. Denoting these species by *T, A, O* and *C*, Cassie found the following correlations with the sample size $n = 50$.

	T	A	O	C
T	1·00	0·74	0·10	0·50
A	0·74	1·00	−0·03	0·15
O	0·10	−0·03	1·00	0·30
C	0·50	0·15	0·30	1·00

This array of correlation coefficients is known as a *correlation matrix*. This matrix is symmetrical since the correlation between Y and X is the same as the correlation between X and Y.

(a) Test at the 1% level the significance of the observed correlation coefficient between T and A counts.

(b) Find the 99% confidence interval for the correlation coefficient between T and C.

(c) In another independent sample with $n = 48$, Cassie found the correlation between T and A counts to be 0.48. Use the Z transformation to test at the 5% level if this coefficient is different from the 0.74 found above.

(a) $H_0: \rho = 0$; $H_1: \rho \neq 0$; $\alpha = 0.01$. Here

$$Z = 0.5 \ln[(1+r)/(1-r)]$$
$$= 0.5 \ln[(1+0.74)/(1-0.74)]$$
$$= 0.950 \qquad \text{from Table 10.1.}$$
$$\sqrt{(n-3)} = \sqrt{47} = 6.856.$$

The observed z is $0.95 \times 6.856 = 6.513$ which is greater than 2.58. Thus $r = 0.74$ is highly significant.

(b) Here

$$Z = 0.5 \ln[(1+0.50)/(1-0.50)]$$
$$= 0.549 \qquad \text{from Table 10.1.}$$
$$\sigma_z = 1/\sqrt{47} = 0.146,$$
$$\text{Lower limit} = 0.549 - 2.58 \times 0.146 = 0.172,$$
$$\text{Upper limit} = 0.549 + 2.58 \times 0.146 = 0.926.$$

These are in the transformed (Z) scale and have to be retransformed. Table 10.1 shows that for $Z = 0.172$, r is 0.17 and for $Z = 0.926$, r is 0.73. The 99% confidence limits are 0.17 and 0.73. As stated in section 10.7, these are not symmetrically placed around the observed r of 0.50.

(c) $H_0: \rho_1 = \rho_2$; $H_1: \rho_1 \neq \rho_2$; $\alpha = 0.05$.

$$n_1 = 50, \qquad\qquad n_2 = 48,$$
$$r_1 = 0.74, \qquad\qquad r_2 = 0.48,$$
$$Z_1 = 0.95, \qquad\qquad Z_2 = 0.52,$$
$$1/(n_1 - 3) = 0.02128, \qquad 1/(n_2 - 3) = 0.02222,$$
$$\text{var}(Z_1 - Z_2) = 0.02128 + 0.02222$$
$$= 0.04350.$$
$$\text{Observed } z = (0.95 - 0.52)/\sqrt{0.04350}$$
$$= 2.06.$$

This is greater than 1·96 and hence the difference is significant at the 5% level.

Exercises

10.4   The weights and heights for a sample of twenty-four university students are presented in the following table. For this sample find the mean height, the mean weight, the standard error of the mean height, the standard error of the mean weight, and the correlation of height and weight, and test the significance of this correlation by the analysis of variance.

Weight (lb)	139	128	121	204	192	123	140	136
Height (in)	66	60	61	74	72	62	65	63

Weight (lb)	136	146	195	171	200	189	130	142
Height (in)	63	65	72	70	72	69·	61	67

Weight (lb)	192	174	156	125	188	183	140	137
Height (in)	72	67	64	61	70	70	61	61

10.5   Tchan *et al.* (1961), in studying a method for the estimation of the nutrient status of N and P in soil, considered the correlation between the plant response (Y) and algal response (X) for random samples of soils. For each nutrient draw a scatter diagram, calculate the linear regression of Y on X and the correlation coefficient, and test the hypothesis that $\rho = 0$.

Nutrient: P

Plant Response	25	72	72	67	104	36	42	81	56
Algal Response	24	94	56	63	98	28	40	64	60

Nutrient: N

Plant Response	26	36	16	40	50	48	56
Algal Response	33	39	12	52	52	62	46

Plant Response	106	65	38	43	21	15	38
Algal Response	80	73	46	52	15	24	56

10.6   The following table relates to cotton. It gives the mean soil temperature, germination time (number of days between sowing and the appearance above ground of 80% of seed sown) and mean length of tap root after fourteen days on ten experimental sites.

Mean soil temperature (°F)	84	81	78	77	73	69	64	63	61	60
Germination time (days)	4	5	7	8	9	11	13	14	14	15
Length of taproot (mm)	135	129	115	104	97	92	80	76	75	70

(a) Plot germination time (G) against mean soil temperature (T). Calculate the regression line of G on T and test the significance of this regression by means of the analysis of variance. What is the standard error of the regression coefficient? Find the 95% confidence limits for the regression coefficient. What is the proportion of variation in germination time attributable to soil temperature? Calculate the 99% confidence interval for the correlation coefficient.

(b) Plot length of taproot (L) against mean soil temperature (T). Calculate the regression line of L on T. What is the standard error of the regression coefficient? Test the hypothesis that $\beta = 3$ mm/°F. If you were to calculate the 95% confidence interval would this interval contain 3 mm/°F? Give a reason for your answer. Finally test the hypothesis that $\rho = 0$ (using the analysis of variance) and find the coefficient of determination.

10.7   Referring to Example 10.4(c), Cassie found in the sample of size $n = 48$, a correlation between T and O of 0.46. Is this significantly different from the correlation of 0.10 found in the sample with $n = 50$?

REFERENCES

CASSIE, R. M. (1959). 'Some Correlations in Replicate Plankton Samples', *N.Z. J. Sci.*, **2**, 473–484.

DAVID, F. N. (1938). *Tables of the Ordinates and Probability Integral of the Distribution of the Correlation Coefficient in Small Samples*. Cambridge University Press, Cambridge.

FISHER. R. A. (1921). 'On the 'Probable Error' of a Coefficient of Correlation Deduced from a Small Sample', *Metron*. **1**, 3–32.

FISHER, R. A. and YATES, F. (1963). *Statistical Tables for Biological Agricultural and Medical Research*. Oliver and Boyd, Edinburgh.

TCHAN, Y. T., BALAAM, L. N., HAWKES, R., and DRAETTE, F. (1961). 'Study of Soil Algae. IV. Estimation of the Nutrient Status of Soil Using an Algal Growth Method with Special Reference to Nitrogen and Phosphorus'. *Plant and Soil.* XIV, **2**, 147–158.

COLLATERAL READING

GOULDEN, C. H. (1952). *Methods of Statistical Analysis* (2nd edition). John Wiley, New York. Chapters 6 and 7.

MACK, S. F. (1960). *Elementary Statistics.* Holt, Rinehart and Winston, New York, Chapter 10.

PARADINE, C. G. and RIVETT, B. H. P. (1969). *Statistical Methods for Technologists* (2nd edition). English Universities Press, London. Chapter 11.

SNEDECOR, G. W. and COCHRAN, W. G. (1967). *Statistical Methods* (6th edition). Iowa State University Press, Ames, Iowa. Chapters 6 and 7.

STEEL, R. G. D. and TORRIE, J. H. (1960). *Principles and Procedures of Statistics.* McGraw-Hill, New York. Chapters 9 and 10.

The χ^2 Test of Goodness of Fit

11.1 The Test Criterion

One of the many problems in biology which involve discrete variables is considered in this chapter. Here the characteristics to be studied are qualitative rather than quantitative and the data consist of the numbers of individuals which fall into well defined classes. For example, a geneticist studying the numbers of individuals in segregating populations desires to test whether the observed numbers are significantly different from those which are expected on the basis of a genetic hypothesis being examined. In problems of this type the χ^2 distribution, which has been defined in Chapter 5 in terms of continuous variables, is used. However since discrete data are now being considered, the test criterion which is developed is not a χ^2 variate but is distributed only approximately as χ^2. In most situations which the biologist has to face, this approximation is very close and the usual χ^2 tables can be used.

Suppose that a total of N individuals can be classified in k classes and that the observed numbers of individuals in these classes are $O_1, O_2, \ldots, O_i, \ldots, O_k$ so that

$$\sum_{i=1}^{k} O_i = N.$$

It should be noted that the O_i are observed frequencies (actual numbers) and are never percentages, and also that $k < N$.

Suppose now that on the basis of a hypothesis being tested, the expected frequencies in the k classes are $E_1, E_2, \ldots, E_i, \ldots, E_k$. The hypothesis might, for example, be that the N observed values are a sample from an infinite normal population, or again that the variate being studied follows a Poisson distribution, or again that the sample is from a population segregating in a $9:3:3:1$ ratio.

The data to be analyzed may be presented as in Table 11.1 In addition to the columns of observed and expected frequencies, there is a column of differences, $O_i - E_i$. For each of the k classes, the difference between the observed number and expected number is calculated. Because of the way in which the E_i are computed,

$$\sum E_i = N$$

and hence

$$\sum (O_i - E_i) = 0.$$

Thus the total of the fourth column is zero.

TABLE 11.1   *Observed and expected frequencies*

Class	Observed frequency	Expected frequency	Difference in frequencies
1	O_1	E_1	$O_1 - E_1$
2	O_2	E_2	$O_2 - E_2$
.	.	.	.
.	.	.	.
i	O_i	E_i	$O_i - E_i$
.	.	.	.
.	.	.	.
k	O_k	E_k	$O_k - E_k$
Column Total	N	N	0

The discrepancies between the O_i and E_i are $O_i - E_i$, and in the testing of the hypothesis on which the expected frequencies were calculated, answers to the following questions are required: 'Are these discrepancies, considered as a whole, large or small? If the hypothesis holds, what is the probability of getting due to chance, a set of O_i which diverges from the expected set of E_i by as much as, or more than, the observed set of O_i?'

The solution to this problem came with the discovery of the χ^2 distribution by Karl Pearson (1900). Its application to problems of the type to be considered in this chapter, where various numbers of classes are involved and where there are various linear restrictions on the calculation of the expected frequencies, was clarified by Fisher (1948) who indicated clearly the need for using the degrees of freedom appropriate to the test criterion.

The test criterion used is

$$\chi^2 = \sum_{i=1}^{k} \frac{(O_i - E_i)^2}{E_i}$$

$$= \sum \frac{(O - E)^2}{E} \qquad \text{(briefly)}.$$

In computing χ^2 the deviations $(O_i - E_i)$ are squared, divided by E_i to approximately equalize the variances and then added. The sum is taken over the k classes. Examination of the test criterion shows that in general a small value of χ^2 reflects small relative discrepancies between the observed and expected frequencies and indicates a good fit. The bigger the overall relative differences between observed and expected frequencies, the larger is χ^2.

To show that $\sum (O - E)^2/E$ is approximately distributed as a χ^2 requires mathematics above the level of this text, and for this reason a rule of thumb by which the number of degrees of freedom is to be calculated will have to be given. This rule is important since using the wrong number of degrees of freedom is one of the ways in which the χ^2 test of goodness of fit is sometimes misused. Other mistakes which are made in the use of χ^2, such as using data that are not observed numbers, are given by Lewis and Burke (1949).

For samples of fixed size the number of degrees of freedom is less than the number of classes, since the sum of the observed and expected frequencies must be the same. This agreement in totals is spoken of as a linear constraint and the number of degrees of freedom if this were the only linear constraint would be the number of classes less one. This is often the only rule that some scientists have applied in the past when using χ^2 tests of goodness of fit. However, there are many occasions when the degrees of freedom are fewer than one less than the number of classes. For instance, to determine the expected frequencies it is often necessary to estimate parameters of the population from the data of the sample. Each estimate of a parameter obtained in this manner corresponds to the introduction of a linear constraint.

As a rule of thumb,

Number of degrees of freedom (ν) for χ^2
= Number of classes − Number of constraints.

An example of when a parameter would have to be estimated is the test of agreement with a Poisson distribution when the parameter λ is not specified. In this case the sample mean would be used to estimate λ and to calculate the expected frequencies. The observed and expected frequency distributions would have the same mean. The degrees of freedom in this test would be two less than the number of classes since there are two linear constraints (agreement of totals and agreement of means).

Another example is the testing of the hypothesis that the sample is a random one from a normal population with unspecified parameters. Both the mean and standard deviation have to be estimated from the sample and are used to find the expected frequencies. Here the number of degrees of freedom is three less than the number of classes.

The χ^2 test of goodness of fit is completed by entering the χ^2 table with the appropriate degrees of freedom and obtaining the probability of getting a value of χ^2 as large as, or larger than, the observed value. If this probability is small (0·05 or 0·01) it is concluded that the fit is not good. As in the standard test of significance procedure, the value of χ^2 is said to be significantly, or highly significantly, large and the hypothesis being tested is rejected.

Just as the analysis of variance was based on certain assumptions and a significant F could be interpreted as rejection of equality of means only if these assumptions were satisfied, the χ^2 test is based on assumptions such as independence and random sampling. These need to be considered when rejecting the null hypothesis and it should not be forgotten that a significantly large χ^2 can be the result of the sampling procedure not yielding a random sample. It might also occur because of differential viability in different classes when a geneticist is testing segregation ratios.

It should also be noted that the calculated value of χ^2 may be significantly small. This would indicate unusually good agreement between the observed and expected distributions, and repeated instances of significantly small values of χ^2 could indicate either over-ingenuity in the selection of the hypothesis or a prepared sample of observations, i.e. it might be suspected that the data are 'cooked'.

An instance of such 'cooking' was a class of genetics students who as part of a laboratory session were required to examine 200 pollen grains under a microscope and classify them into two classes. All students being good at elementary genetical theory knew that the expected ratio was 1:1. However, a number of students, being neither good at biometry nor willing to examine 200 grains under a microscope, elected to short-cut the exercise and 'made a stab' at the final numbers in the two genetic classes. Another small group of students, who were more conscientious than their colleagues noted after counting 200 grains that their observed numbers showed to their way of thinking considerable deviation from expectation, and continued counting until their observed ratio agreed very well with the expected. Both types of error were detected and admitted when about one quarter of the class were found to have significantly small values of χ^2. ('Lies, damned lies, statistics'—Disraeli.)

Another point to note before considering some examples is that the mathematical derivation of the χ^2 test requires that all the observed and expected frequencies shall be 'sufficiently large'. For this reason, Fisher (1948) suggested that the number expected in any group be not less than five, and that when some of the frequencies are below this minimum it may be necessary to group two or more adjacent classes into one class. Later Cochran (1954) showed that this was too conservative and recommended grouping so that the minimum expected frequency be at least one in any class.

Example 11.1 In studying segregation for seedling mildew reaction in the progenies of F_1 plants from crosses between resistant and susceptible wheat varieties, the following data were obtained:

Number of resistant plants	215
Number of susceptible plants	85
Total	300.

Test by two methods the hypothesis that the ratio of resistant to susceptible is 3:1.

Method (i)   In the first method the χ^2 test of goodness of fit is used.

Null hypothesis, H_0:   Ratio is 3:1,
Alternative hypothesis, H_1:   Ratio is different from 3:1,
Level of significance (α):   0·05.

On the basis of the null hypothesis, the expected frequencies for the two classes are respectively $\frac{3}{4} \times 300 = 225$ and $\frac{1}{4} \times 300 = 75$.

Table 11.2 shows the setting out of the data and calculation of χ^2.

TABLE 11.2 *Calculation of* χ^2

Class	Observed frequency (O_i)	Expected frequency (E_i)	$O_i - E_i$	$(O_i - E_i)^2/E_i$
Resistant	215	225	-10	0·444
Susceptible	85	75	$+10$	1·333
Total	300	300	0·00	1·777

In this example, the expected values are calculated from the observed total so that only one class can be filled arbitrarily, for once a frequency is assigned to the resistant class the number in the susceptible class is determined (it is the total frequency minus the frequency of the resistant class). Thus χ^2 has one degree of freedom.

From the table of χ^2 the probability (p) of obtaining a χ^2 as large as, or larger than, 1·777 is between 0·1 and 0·2.

Thus $p(\doteqdot 0{\cdot}18) \not< \alpha(= 0{\cdot}05)$ and the null hypothesis is not rejected. There is insufficient evidence to conclude that the ratio is different from 3:1.

Method (ii) As stated in section 5.4, χ^2 with one degree of freedom is the square of a standardized normal variate. This fact is used in this second method which is equivalent to the first.

Let π be the probability of a plant being resistant.

$$H_0 : \pi = \tfrac{3}{4},$$
$$H_1 : \pi \neq \tfrac{3}{4},$$

With $\pi = \tfrac{3}{4}$, the probabilities of groups of 300 plants having 300, 299, 298, etc., resistant plants could be found using the terms of the binomial distribution by the expansion of $(\tfrac{3}{4} + \tfrac{1}{4})^{300}$. From a knowledge of these individual terms an exact test of significance could be carried out. However this approach would be very time consuming.

However since $n = 300$ is large, the knowledge that the binomial distribution approximates closely the normal distribution when n is large can be used.

With $\pi = \frac{3}{4}$, the expected number (E) of resistant plants is 225. The observed number (O) is 215. The standard deviation (σ) of the distribution of observed number for a binomial distribution with parameters n and π is

$$\sigma = \sqrt{(n\pi\delta)} \qquad \text{where } \delta = 1 - \pi$$
$$= \sqrt{(300 \times \tfrac{3}{4} \times \tfrac{1}{4})}$$
$$= 7\cdot5.$$

$(O - E)/\sigma$ is a standardized normal variate (approximation is very good) and in this example

$$(O - E)/\sigma = -10/7\cdot5 = -1\cdot333.$$

Now a value $< -1\cdot333$ or $> 1\cdot333$ of a standardized normal variate is expected by chance about once in 5 or 6 trials, or in other words, the observed p is about $0\cdot18$. Thus

$$p(\doteqdot 0\cdot18) \nless \alpha(= 0\cdot05)$$

and the null hypothesis is not rejected.

The two methods give the same result, as indeed they must if the tests are suitable and the calculations done correctly in both instances. In passing it should be noted that

$$\left[\frac{O - E}{\sigma}\right]^2 = (-1\cdot333)^2 = 1\cdot777 = \chi^2_{1\text{d.f.}}.$$

This second method is of use and is easy to apply to such cases as the above where there are only two classes. It is not easily adapted to problems where there are several classes.

Example 11.2 In a certain cross the types represented by BC, Bc, bC, bc are expected to occur in a $9\!:\!3\!:\!3\!:\!1$ ratio. The observed frequencies were

BC	Bc	bC	bc
328	122	77	33

Are these observations in accordance with this hypothesis?

Null hypothesis, H_0: Ratio $9\!:\!3\!:\!3\!:\!1$,
Alternative hypothesis, H_1: The ratio is different from $9\!:\!3\!:\!3\!:\!1$,
Level of significance (α): $0\cdot05$.

The calculations for χ^2 are shown in Table 11.3.

Class	Observed frequency (O_i)	Expected frequency (E_i)	$O_i - E_i$	$(O_i - E_i)^2/E_i$
BC	328	315	13	0·537
Bc	122	105	17	2·752
bC	77	105	−28	7·467
bc	33	35	− 2	0·114
	560	560	0	10·87

If the ratio is 9:3:3:1 the expected proportions in the four classes are $\frac{9}{16}, \frac{3}{16}, \frac{3}{16}, \frac{1}{16}$

The expected values are calculated from the observed total, e.g.

expected frequency for $BC = \frac{9}{16} \times 560 = 315$, etc.,

so that the four observed and expected classes must agree in their totals. Thus

No. of degrees of freedom = No. of classes − No. of constraints

$$= 4 - 1 = 3.$$

Hence $v = 3$ and the probability of obtaining a χ^2 as large as, or larger than, 10·87 is between 0·01 and 0·02.

Thus $p(\doteqdot 0·02) < \alpha(= 0·05)$ and the observed value of χ^2 is significant at the 5% level. The null hypothesis is rejected and it is concluded that the ratio is different from 9:3:3:1.

Example 11.3 Table 11.4 gives the distribution of yields per plot in grams for 294 plots in a wheat uniformity trial.

TABLE 11.4   *Frequency distribution of yields in grams per plot*

Class limits	Frequency	Class limits	Frequency
0·05–10·05	8	50·05–60·05	54
10·05–20·05	20	60·05–70·05	34
20·05–30·05	35	70·05–80·05	12
30·05–40·05	53	80·05–90·05	9
40·05–50·05	69		

Use the χ^2 test of goodness of fit to test whether the results are consistent with the hypothesis that yield per plot is normally distributed.

To obtain the expected frequencies (E_i), a normal curve has to be fitted to the data. In fitting a normal curve to the observed data, the observed frequencies (O_i) are used to calculate the mean

$$\bar{x}_. = \sum O_i X_i / N$$

and variance

$$s^2 = \frac{\sum O_i X_i^2 - (\sum O_i X_i)^2 / N}{N - 1}$$

where N is the total frequency and X_i is the mid-point of the ith class. Then Table I is used to obtain the proportions of observations expected in the various classes on the assumption that the population is normal with mean equal to $\bar{x}$ and variance equal to s^2. The proportions which sum to unity are finally multiplied by N to give the expected frequencies.

Because the mean and variance of the observed distribution are used in determining the expected frequencies, the expected distribution agrees with the observed in total, mean and variance and thus there are 3 constraints. Then since there are 9 classes and 3 constraints χ^2 has $6(= 9 - 3)$ degrees of freedom.

The expected frequencies are

$$9{\cdot}1, 17{\cdot}9, 37{\cdot}0, 55{\cdot}3, 62{\cdot}6, 54{\cdot}1, 33{\cdot}8, 16{\cdot}3, 7{\cdot}9$$

and

$$\chi^2 = \frac{(1{\cdot}1)^2}{9{\cdot}1} + \frac{(2{\cdot}1)^2}{17{\cdot}9} + \ldots + \frac{(1{\cdot}1)^2}{7{\cdot}9} = 2{\cdot}48.$$

From the χ^2 table, with $v = 6$, the probability (p) of $2{\cdot}48$ being exceeded in random sampling lies between $0{\cdot}90$ and $0{\cdot}80$. This observed probability is quite large and not even close to the 5% level of significance. There is little or no evidence to reject the null hypothesis that wheat yields in the population, from which this sample was drawn, are normally distributed.

Exercises

11.1 The following is a summary of F_2 results of Mendel's (1866) experiments with peas. Test the hypothesis of a $3:1$ ratio for the data on all characters by two methods. (First method—approximation of binomial to normal; second method—χ^2 test of goodness of fit.)

Character	Dominant	Recessive	Total
Colour of cotyledons	6022	2001	8023
Form of seed	5474	1850	7324
Form of pod	882	299	1181
Length of stem	787	277	1064
Colour of seed coats and flowers	705	224	929
Position of flowers	651	207	858
Colour of pod	428	152	580

11.2 The data given below represent segregations in five F_2 populations. Test the goodness of fit of each population to a $9:3:3:1$ ratio, and determine which may be regarded as examples of this ratio and which may not.

Population	AB	Ab	aB	ab
1	72	43	39	6
2	864	317	341	118
3	1770	610	618	202
4	75	35	41	9
5	51	11	16	2

11.3 By calculating values of χ^2, determine whether the following F_2 population fits better a $3:3:1:1$ ratio or a $27:21:9:7$ ratio.

BC	Bc	bC	bc
288	227	96	93.

11.4 The following table gives the number of yeast cells in 400 squares of a haemacytometer.

Number of cells	0	1	2	3	4	5	Total
Frequency	213	128	37	18	3	1	400

Using the mean of the observed distribution ($\bar{x} = 0.68$), a Poisson distribution is fitted.

Expected frequency: 203   138   47   11   1   0

Again using the mean and variance of the observed distribution, a negative binomial distribution is fitted.

Expected frequency: 214   123   45   13   4   1

Apply the χ^2 test of goodness of fit to test the hypothesis that the distribution of yeast cells is (a) Poisson and (b) negative binomial.

11.5   Mendel (1866) in his classic genetic study, observed plant to plant variation in an experiment on the colour of the seed albumen. For the first 10 plants the results were:

Plant	1	2	3	4	5	6	7	8	9	10
Yellow albumen	25	32	14	70	24	20	32	44	50	44
Green albumen	11	7	5	27	13	6	13	9	14	18

Compute values of χ^2 for each plant to test H_0: 3:1 ratio. Find the sum of these χ^2s. How many degrees of freedom will this total χ^2 have? (Refer back to section 6.12.)

11.6   In a cross of two wheat varieties, the following results with regard to frost resistance were obtained:

Resistant	20
Semi-resistant	38
Susceptible	86
	——
	144

Two hypotheses are available according to which the expected proportions are:
First hypothesis—
   resistant: semi-resistant: susceptible = 1 :2 :3.
Second hypothesis—
   resistant: semi-resistant: susceptible = 1 :2 :5.
(a) Use the χ^2 test to investigate the two hypotheses.
(b) What conclusions would you draw if instead of the sample of 144, a sample 10 times as large had been available with the same proportions of resistant, semi-resistant and susceptible plants?

11.7 In a blood count of 1000 squares it was found that:

> 625 squares contained no blood cells;
> 275 squares contained one blood cell;
> 80 squares contained two blood cells;
> 15 squares contained three blood cells;
> 5 squares contained four blood cells.

Use the mean of the observed distribution to calculate the expected frequencies for the various classes on the assumption that the number of blood cells in a random square is a Poisson variate. Test this assumption.

REFERENCES

COCHRAN, W. G. (1954). 'Some Methods for Strengthening the Common χ^2 tests', *Biometrics*, **10**, 417–451.

FISHER, R. A. (1948, 1950). *Statistical Methods for Research Workers* (10th, 11th editions). Oliver & Boyd, Edinburgh; Hafner, New York.

LEWIS, D. and BURKE, C. J. (1949). 'The Use and Misuse of the Chi-square Test', *Psych. Bull.*, **46**, 433–498.

MENDEL, G. (1866). 'Versuche über Pflanzenhybriden', (Experiments in Plant Hybridization), English translations from the Harvard University Press, Cambridge, Mass. (1948); Oliver & Boyd, Edinburgh, (1965).

PEARSON, K. (1900). 'On a Criterion that a Given System of Deviations from the Probable in the Case of a Correlated System of Variables is such that it Can be Reasonably Supposed to Have Arisen in Random Sampling', *Phil. Mag.*, 5, 1, 157–175.

COLLATERAL READING

BLISS, C. I. (1967). *Statistics in Biology*. McGraw-Hill, New York. Chapter 3.

GOULDEN, C. H. (1952). *Methods of Statistical Analysis*. John Wiley. New York. Chapter 15.

MATHER, K. (1951). *The Measurement of Linkage in Heredity*. Methuen, London. Chapter 2.

PARADINE, C. G. and RIVETT, B. H. P. (1969). *Statistical Methods for Technologists* (2nd edition). English Universities Press, London. Chapter 5.

SOKAL, R. R. and ROHLF, F. J. (1969). *Biometry*, W. H. Freeman, San Francisco. Chapter 16.

CHAPTER 12

Some Non-parametric Tests

12.1 Non-Parametric v. Parametric Statistics

The statistical techniques which have been developed in the previous chapters have considered the estimation of parameters and the testing of hypotheses concerning them. These are known as *parametric* statistics. Models have been used which specified certain conditions about the parameters of the population or populations from which the sample or samples were drawn. In this chapter some non-parametric tests are considered. These are important and are being used more and more nowadays because the assumptions concerning the model are fewer and not as strong as those for the parametric tests (z, t, F). In many of the tests which have been considered, an underlying assumption of normality has been required. In certain areas of biological research this assumption is only partly fulfilled. If the underlying distribution is known then transformations may be made. However, there are many instances in which the underlying distribution is not known, and it is in these cases that non-parametric methods are useful.

The development of non-parametric statistics arose because of the need for statistical methods which had desirable properties when little was assumed known about the population (or populations) from which the sample or samples were drawn. While non-parametric statistics, according to Bradley (1960), can be traced back as far as 1710, the real development in the subject has taken place from about the middle 1930s. The growth and interest in the subject since then has been considerable, and while it may be regarded as unfortunate that some scientists use non-parametric procedures all the time, whether they are appropriate or not. it is equally unfortunate that others tend in general to overlook their existence.

12.2   Measurement

Before considering, even at an elementary level, some of the non-parametric procedures, it is necessary to understand the rudiments of the theory of measurement. This theory determines the operations which may be performed on a variate. The allowable operations depend on the level of measurement. Following Siegel (1956) four levels of measurement are defined.

The weakest or lowest level is when the *nominal scale* is used. Here individuals are measured simply by a word or a number. The variable is frequently a qualitative one—dead or alive; awned or awnless; urban, semi-rural, rural. Numbers are often used to classify or identify individuals—e.g. student identification numbers; plant-breeding strain numbers. At this level of measurement, the scale is used to divide the whole group of individuals into two (or more) sub-groups. While this is the lowest level of measurement, it is often the only one which biologists can use.

The next level of measurement is the *ordinal* or *ranking scale*. Here the individuals in one category of the scale are not just different from those in another category—they stand in some kind of relation to each other, e.g. > or <. Examples are

(i)   food technologists using palatability scales—0: very unpalatable, 1: unpalatable, ..., 5: most palatable;

(ii) agronomists scoring root or plant damage—0: no damage, ..., 4: very badly damaged;

(iii) plant breeders classifying plants as short, medium, tall;

(iv) geneticists studying flower colour on a colour scale.

The third level of measurement is the *interval scale* in which distances between two units on the scale are known. For each interval scale there is a zero point and a common constant unit of measurement; both of which are arbitrarily chosen. The ratio of any two intervals is independent of the unit of measurement. A typical and frequently quoted example of interval scales is the Fahrenheit and Celsius (formerly centigrade) temperature scales. Four points on these two scales are

| Fahrenheit | 32 | 68 | 122 | 212 |
| Celsius | 0 | 20 | 50 | 100 |

The ratio of the difference between the third and second points on the Fahrenheit scale ($122°-68° = 54°$) to the difference between the fourth and second ($212°-68° = 144°$) is 0.375. On the Celsius scale the ratio of these two intervals ($30°:80°$) is also 0.375. This equality holds for all pairs of intervals and is a characteristic of an interval scale. However, the ratio of the third point to the second on the Fahrenheit scale is 1.79, the ratio of the third to the second on the Celsius scale is 2.5. Only the ratios of intervals are equal.

The fourth and highest level of measurement is the *ratio scale*. This is an interval scale with a true zero point. Examples are measurements of mass and length. The ratio of the girths of two animals would be the same irrespective of whether the measurement was made in inches or in centimetres.

12.3 Advantages and Disadvantages

Both non-parametric and parametric procedures have their advantages and disadvantages. Because they extract as much information from the sample as possible, parametric procedures will always be superior to non-parametric procedures provided the data are such that the parametric procedure may rightly be used. The assumptions on which the parametric procedure is based must be satisfied. Those of normality, equality of variance, independence, etc., have been stressed in the earlier chapters. From the previous section it will be apparent that parametric procedures require data on at least the interval scale. Not only ranking but the size of the difference between the values of the variable must be known.

If the parametric procedure assumptions are not satisfied, then non-parametric statistics have a number of advantages. The first of these which might appeal to the research worker not well trained in mathematics or statistics is that the derivations of non-parametric tests are usually more easily understood. This is because the tests can frequently be derived using simple combinatorial formulae. Thus the researcher is less likely to feel that he is only following a recipe book. Another advantage is that the calculation of the non-parametric test criterion is usually relatively simple. Frequently the arithmetical calculations can be carried out using only pencil and paper, whereas most parametric tests require a calculating machine. Further the assumptions on which non-parametric procedures are based are fewer than for parametric procedures. Most of the tests can be applied to data on an ordinal scale, i.e. to ranked data. Frequently nothing more is assumed about the parent distribution than that it be continuous. For this reason non-parametric statistics have come to be known as 'distribution-free'. For example when the sign test (illustrated later in this chapter) is applied to a variate Y, the null hypothesis is that Y has the same distribution under two treatments. However it is not necessary in setting up the null hypothesis to state the form of the distribution. This contrasts with the t test where the null hypothesis specifies normal homoscedastic distributions and a relationship (frequently equality) between the two population parameters, μ_i. The t test is a *parametric* test whereas the sign test is *non-parametric*. Instead of testing means, non-parametric procedures often test medians which are non-parametric estimates for any continuous distribution. Finally, some non-parametric tests such as the Kolmogorov-Smirnov test are used when it is desired to test for any difference whatsoever between two populations.

12.4 The t Test based on Range

As indicated in section 2.6, the range (difference between largest and smallest member) is used as a measure of spread. The sample range (just as the sample standard deviation) is an estimator of the population standard deviation. Lord (1947) uses the sample range (w) in an alternative to the t test to make tests of significance concerning population means. As in the case of the t distribution, confidence interval estimates can also be made. An underlying assumption of Lord's procedure is normality and while his test is not a non-parametric test it is introduced at this point because it possesses one of the advantages of non-parametric methods. It is quick and easy to apply.

Lord's test criterion, for which he presents critical points for one-tailed and two-tailed tests of significance, is

$$t_w = \frac{\bar{x}. - \mu_0}{w} \text{ for a null hypothesis } H_0 : \mu = \mu_0,$$

or

$$t_w = \frac{\bar{x}_{1.} - \bar{x}_{2.}}{\frac{1}{2}(w_1 + w_2)}$$

$$= \frac{\bar{x}_{1.} - \bar{x}_{2.}}{\bar{w}} \text{ for a null hypothesis } H_0 : \mu_1 = \mu_2.$$

Data from Example 7.1 are used to illustrate Lord's procedure in the case of a single sample. Here $\bar{x}_. = 13\cdot0$ and $w = 13\cdot9 - 12\cdot4 = 1\cdot5$. To test the hypothesis that $\mu = 12\cdot66$,

$$t_w = (13\cdot0 - 12\cdot66)/1\cdot5 = 0\cdot227.$$

For a sample of size 9, Lord (1947) gives the critical points of t_w as $\pm0\cdot255$. Thus the null hypothesis is not rejected at this level of significance. The 95% confidence interval using Lord's procedure is

$$\bar{x}_. - t_w w \leqslant \mu \leqslant \bar{x}_. + t_w w$$

which is

$$13\cdot0 - 0\cdot255 \times 1\cdot5 \leqslant \mu \leqslant 13\cdot0 + 0\cdot255 \times 1\cdot5.$$

This interval $\{12\cdot62 \text{ to } 13\cdot38\}$ is only slightly different from the interval $\{12\cdot60 \text{ to } 13\cdot40\}$ obtained with the t distribution.

In the case of two means the method is illustrated using the data in Example 7.2. Here

$$\bar{x}_{1.} = 26\cdot75,$$
$$\bar{x}_{2.} = 21\cdot99,$$
$$w_1 = 37\cdot0 - 19\cdot8 = 17\cdot2,$$
$$w_2 = 34\cdot0 - 13\cdot7 = 20\cdot3,$$
$$\bar{w} = 18\cdot75,$$
$$t_w = (26\cdot75 - 21\cdot98)/18\cdot75 = 0\cdot254.$$

From tables presented by Lord this value is not significant at the 5% level.

Just as the t test can be extended to those cases where the sample sizes are different, Lord's procedure may also be extended. Moore (1957) has presented tables for unequal sample sizes. The critical points are tabulated where the sizes of both samples are 20 or less.

For the case of 'paired comparisons', the test statistic is $t_w = \bar{d}_. / w_d$

where w_d is the range in the sample of differences. Here Lord's table has to be entered with the number of pairs or differences which are involved.

This test based on the range is nearly as efficient as the t-test when the sample size is small and where many tests have to be undertaken. Biologists might well consider its use where a small, constant number of items or experimental units are examined routinely at regular intervals. For example, in examining the weight of eggs packaged for marketing, samples of size one dozen might be taken from every twentieth case and the mean weight tested using the above test.

12.5 Contingency Tables

Biologists using data on a nominal scale frequently present their observations in a contingency table which is composed of r rows and c columns. For example, plant physiologists studying the number of seeds which germinate under different treatments, geneticists investigating egg hatchability of different strains of fruit fly, or plant breeders considering rust reaction in the seedling stage and in the field, all have occasion to use contingency tables. In section 1.2 a simple 2×2 contingency table was presented. Now if it were desired to test the hypothesis that the dead: alive classification is independent of the insecticide which was applied, the following test could be made. This test holds irrespective of the number of rows or columns in the contingency table.

The agreement between the observed and expected frequencies is studied by means of the χ^2 distribution. The expected frequencies $(E_{ij}; i = 1, 2, \ldots, r; j = 1, 2, \ldots, c)$ in a table with r rows and c columns can be calculated using the marginal frequencies if independence is assumed. They are calculated from the formula

$$E_{ij} = \frac{n_{i.} n_{.j}}{n_{..}}$$

where $n_{i.}$ are the marginal row totals,

$n_{.j}$ are the marginal column totals,

$n_{..}$ is the total number of individuals classified in the table. This formula is developed using the theorem of compound probability. On the assumption of independence the proportion of individuals expected in the (i,j)th cell is $p_{i.} \times p_{.j}$ where

$$p_{i.} = n_{i.}/n_{..} \text{ and } p_{.j} = n_{.j}/n_{..}.$$

Hence the expected number of individuals in the (i,j)th cell is $p_{i.} \times p_{.j} \times n_{..}$ which is $n_{i.} \times n_{.j}/n_{..}$.

The χ^2 test criterion is $\sum (O_i - E_i)^2/E_i$. For a contingency table with r rows and c columns, this χ^2 has $(r-1)(c-1)$ degrees of freedom because the row and column totals in the tables of observed and expected frequencies agree. For a 2×2 contingency table as in section 1.2, the χ^2 has 1 degree of freedom. Once the expected frequency for one of the cells is calculated using the marginal observed frequencies, the three remaining expected frequencies are determined. This is so because the expected and marginal totals agree.

The χ^2 test used here is a non-parametric test because the expected frequencies are not based on parameters.

12.6   The Kolmogorov–Smirnov Test

In Example 11.3 the χ^2 test of goodness of fit was used to test whether data in a frequency distribution were consistent with the hypothesis that the data were a sample from a normal population. Again, in Exercise 11.4 the χ^2 test was used to test for a Poisson distribution and a negative binomial. The test which is introduced here is an alternative to that used in Example 11.3 where the variable, yield, whose distribution was being investigated is continuous. It is not appropriate where the variable is discrete, as in Exercise 11.4.

The non-parametric test for goodness of fit which is to be considered in this section is known as the Kolmogorov–Smirnov test. It was first suggested and studied by Kolmogorov (1933) and Smirnov (1948). The test detects departures from the null hypothesis better than does the corresponding χ^2 test and as is the case with most non-parametric tests it is relatively simpler to perform.

In the Kolmogorov–Smirnov test, the observed and expected cumulative frequency distributions are compared and the point at which these two distributions show the greatest divergence is determined. This divergence forms the basis for the test. Suppose $F(X)$ defines the cumulative frequency distribution under the null hypothesis; $F(X)$ is the expected number of observations equal to or less than X. Suppose further that $O_n(X)$ defines the observed cumulative frequency distribution for the sample of size n. If f_i is the observed frequency for $X = X_i$ and if $X_1 \leqslant X_2 \leqslant X_3 \leqslant \ldots$, then $\sum_{i=1}^{p} f_i$ is the cumulative frequency to $X = X_p$ and

$$O_n(X_p) = \sum_{i=1}^{p} f_i.$$

The largest difference between $F(X)$ and $O_n(X)$ is

$$d_{max} = \max |F(X) - O_n(X)|.$$

The test statistic is $D = d_{max}/n$.

The sampling distribution of D has been studied by Massey (1951) who has tabulated the distribution for $n \leqslant 35$. This can be used to determine whether a significantly large D has been observed and hence whether the null hypothesis is rejected or not. For $n > 35$, the 5% and 1% critical points of D are $1\cdot36/\sqrt{n}$ and $1\cdot63/\sqrt{n}$.

Example 12.1 Use the Kolmogorov–Smirnov test to test whether the data in Example 11.3 are consistent with the hypothesis that the 294 yields are a random sample from a normal population.

Here the expected frequencies are found using the mean and standard deviation as estimated from the sample. The observed and expected cumulative frequency distributions, using the results obtained in Example 11.3, are presented in Table 12.1. In this table

TABLE 12.1   *Distributions and deviations used in Kolmogorov–Smirnov test*

Upper class limit	Cumulative observed frequency	cumulative expected frequency	d
10·05	8	9·1	1·1
20·05	28	27·0	−1·0
30·05	63	64·0	−1·0
40·05	116	119·3	3·3
50·05	185	181·9	−3·1
60·05	239	236·0	−3·0
70·05	273	269·8	−3·2
80·05	285	286·1	1·1
∞	294	294·0	0

the upper class limit for the last class is shown as ∞. This is so because implicit in the sample is the information that there were no observations greater than 90·05. Hence there are 9 observations greater than 80·05 and less than ∞. With ∞ as the upper class limit the cumulative expected frequency is 294.

Here $d_{max} = 3·3$ and $D = 3·3/294 = 0·011$. The 5% critical point of D is $1·36/\sqrt{294} = 0·079$. As the observed D is less than this there is insufficient evidence to reject the null hypothesis.

It should be noted that even though parameters of the fitted distribution are estimated before the Kolmogorov–Smirnov test is applied, the test is nevertheless a non-parametric test because the distribution of the D statistic is independent of the normal distribution being hypothesized. In contrast, such a statement could not be made for the t-test where the distribution of the t statistic depends on the particular normal distribution being assumed under the null hypothesis. The t-test is a parametric test.

12.7 The Wilcoxon Two-sample Test: The Mann and Whitney U Test

The non-parametric test about to be considered originated with Wilcoxon and was developed by Mann and Whitney. It may be regarded as the non-parametric counterpart to the unpaired t test. In this case, however, there is no assumption of normality, nor is the null hypothesis related to the population means. The null hypothesis is that the two samples come from populations having the same distribution. The two-tailed test detects differences in 'location' of the two distributions.

In this test the two samples are grouped together and ranked from the lowest to the highest. The ranks form the basis of the test. If two observations are equal each is given the average rank of the two. Then the sum $T(= \sum R_i)$ of the ranks of the smaller sample is found. The Wilcoxon or Mann and Whitney statistic is

$$U = [n_1 n_2 + n_1 (n_1 + 1)/2] - T$$

where n_1 is the size of the smaller sample and n_2 is the size of the larger. This statistic is compared with

$$U' = n_1 n_2 - U$$

and the smaller of U' and U is the test statistic.

Like the Kolmogorov–Smirnov test, the present test is concerned with the agreement between cumulative distributions. However, in contrast with the Kolmogorov–Smirnov test statistic and others which have been used previously in this text, small values of the U statistic imply rejection of the null hypothesis. This will be appreciated by considering two samples which do not overlap. If all the members of the sample of size n_1 are less than those of the other sample, then

$$T = 1+2+3+ \ldots +n_1 = n_1(n_1+1)/2.$$

Hence $U = n_1 n_2$ and $U' = 0$. The smaller of these is zero and this is always the case when there is no overlap whatsoever in the samples.

The probabilities associated with values as small as the observed values of U are to be found in Mann and Whitney (1947) and may be used to complete the test of significance provided $n_2 \leqslant 8$. For $8 < n_2 \leqslant 20$, the tables of Auble (1953) which are reproduced in Siegel (1956) may be used. For $n_2 > 20$ the results of Mann and Whitney (1947) can be used. They showed that the sampling distribution of U approaches the normal distribution with mean $\frac{1}{2}n_1 n_2$ and variance $n_1 n_2(n_1+n_2+1)/12$. Thus to complete the test of significance a z-test can be used where

$$z = \frac{U - \frac{1}{2}n_1 n_2}{\sqrt{(n_1 n_2(n_1+n_2+1)/12)}}.$$

Example 12.2 The data from Exercise 7.10 are used to illustrate this procedure. The low protein diet is represented by L; the high by H. The data when ranked are

Weight	9	11	11	12	13	14	14	15	15	15	17	17
Group	L	L	L	H	H	L	H	L	H	H	H	H
Rank	1	2·5	2·5	4	5	6·5	6·5	9	9	9	11·5	11·5

Here $n_1 = 5$ (number of chickens on L)
and $n_2 = 7$ (number of chickens on H),
$T = 1+2\cdot5+2\cdot5+6\cdot5+9 = 21\cdot5,$
$U = 5\times7+(5\times6)/2-21\cdot5 = 28\cdot5,$
$U' = 35-28\cdot5 = 6\cdot5.$

Examination of Mann and Whitney (1947) gives for $n_2 = 7$, $n_1 = 5$ a probability associated with this U of 0·05.

The null hypothesis that the two samples came from the same population is rejected at the 5 % level.

12.8 The Sign Test

A most simple non-parametric alternative to the paired t test is the sign test. This test is so named because it is based on the signs of the differences between the paired observations. The sample of differences between the n pairs of measurements would be expected to have the same number ($\frac{1}{2}n$) of positive and negative values provided the sample were drawn from a population whose median was zero. This test then is based on the hypothesis that the median difference is zero. However for symmetric distributions the median and the mean coincide.

The number of positive and negative differences is recorded and if the sample of differences is large, a χ^2 test of goodness of fit between the observed numbers and the expected ($\frac{1}{2}n$ positives; $\frac{1}{2}n$ negatives) is performed. For smaller samples, the binomial distribution with $\pi = 0.5$ can be used.

Example 12.3 The data from Example 7.4 are used to illustrate this test. Here $n = 10$ and 8 differences are positive, 2 are negative. The binomial distribution probabilities with $n = 10$ and $\pi = 0.5$ are given in Example 4.1. Reference to this example shows that

$$P(X = 0) = 0.001,$$
$$P(X = 1) = 0.010,$$
$$P(X = 2) = 0.044.$$

Thus the probability of a sample as extreme or more extreme than that observed is

$$0.001 + 0.010 + 0.044 = 0.055.$$

Thus the observed sample is on the borderline of significance if a 5% level of significance is chosen. This result might be compared with the highly significant result obtained in Example 7.4. In the present test, only the signs of the differences have been used and not their magnitude. However, while much of the information concerning the differences is neglected in the sign test, it is nevertheless a quick rapid test which is very useful when the number of differences is 20 or more.

12.9　Non-Parametric Tests for the Completely Random Design

Two non-parametric tests for the completely random design are now considered. The first of these is similar to Wilcoxon's test and was developed by Kruskal and Wallis (1952). In their procedure all the observations from the t treatments are ranked together from the smallest to the largest. As in the Wilcoxon test if there are any ties the average rank is given to those observations which are equal. It should be noted that this test differs from that proposed in the following section for the randomized complete block design in that the test presently under consideration requires all the observations to be ranked whereas in the procedure for the randomized complete block design, observations within blocks are ranked. It will be appreciated that if the total number of observations in the experiment is large, then the ranking of all the observations is not easy and care will need to be taken to avoid errors.

Having ranked all the observations, the sum of the ranks for each treatment is obtained. Suppose R_i is the sum of the ranks for the ith treatment which has been applied to n_i experimental units and that $\sum n_i = N$. The test criterion is

$$H = \frac{12}{N(N+1)} \sum_{i=1}^{t} \frac{R_i^2}{n_i} - 3(N+1)$$

H is distributed approximately as χ^2 with $(t-1)$ degrees of freedom for large samples provided the null hypothesis is true. The null hypothesis is that the t samples come from the same population or populations identical with respect to 'location', and the test assumes that the variable under study has a continuous distribution. For $t = 3$ and sample sizes n_i each less than 5, H is not distributed as χ^2. For problems of this magnitude, the exact distribution tabulated by Kruskal and Wallis (1952) should be used.

The null hypothesis is rejected if a significantly large H is obtained. Corrections to the above formula for H have to be applied if there are ties. These corrections increase H slightly and should be used if the observed H approaches significance.

The second test presented for a completely random design is a median test. In this test all the observations are again grouped together and ranked from smallest to largest. The median is found and, for each treatment the number of observations above and below the overall median is determined. A $2 \times t$ contingency table showing the numbers above and below the median for each treatment is prepared. These observed numbers may then be compared with expected numbers on the assumption that all treatment medians are equal. The contingency χ^2 has $(t-1)$ degrees of freedom.

Example 12.4 Using the data presented in Exercise 9.7, test by the Kruskal–Wallis procedure the hypothesis that the five strains are samples from the same population.
The ranked data are:

Yield	17	18	19	20	20	21	21	21	22	23
Strain	D	D	D	E	B	D	E	B	B	B
Rank	1	2	3	4·5	4·5	7	7	7	9	11

Yield	23	23	24	25	26	26	27	27	28	29
Strain	C	E	A	C	E	A	A	C	C	A
Rank	11	11	13	14	15·5	15·5	17·5	17·5	19	20

The ranks of the five strains are:

Strain A	13	15·5	17·5	20	$R_1 = 66$
Strain B	4·5	7	9	11	$R_2 = 31·5$
Strain C	11	14	17·5	19	$R_3 = 61·5$
Strain D	1	2	3	7	$R_4 = 13$
Strain E	4·5	7	11	15·5	$R_5 = 38$

The test criterion is

$$H = \frac{12}{20 \times 21} \times \frac{10743·5}{4} - 3 \times 21$$
$$= 13·74.$$

On entering the χ^2 table with four degrees of freedom, this observed H is found to be just significant at the 1 % level. Adjustment for the ties is hardly warranted since this would only increase H.

Example 12.5 Five groups of chickens were fed on five different

diets (A, B, C, D, E). The weights in grams at the end of the experiment
are given below.

Diet A

256	292	286	275	314	277	267	294	304	262
297	302	304	320	284	290	268	292	310	262
279	274	282	315	296	321	326	299	305	298

Diet B

304	315	306	319	323	310	313	307	324	309
315	324	332	318	295	273	274	294	331	335
287	342	293	327	321	347	284			

Diet C

284	329	292	314	307	332	315	335	323	319
343	343	359	339	314	306	326	298	334	309
342	317	352	329	357	320				

Diet D

263	285	275	297	287	309	294	313	306	323
315	288	321	291	339	305	275	312	282	343
296	309	304	311	312	327	301	308		

Diet E

230	254	241	265	252	276	263	287	274	298
286	309	296	261	307	272	318	283	249	274
259	276	270	281	292	303				

The overall median is 304. The numbers above and below this
median for each diet are recorded in the following table. The expected
numbers are also presented.

Treatment	Observed numbers		Treatment total	Expected numbers	
	above median	below median		above median	below median
A	21	7	28	14	14
B	19	7	26	13	13
C	23	3	26	13	13
D	15	12	27	13·5	13·5
E	3	23	26	13	13

The observed χ^2 is 43·64 which on entering the χ^2 table with four degrees of freedom is highly significant. The null hypothesis that the five treatment medians are equal is rejected.

12.10   The Randomized Complete Block: Friedman's Procedure

A non-parametric test for the randomized complete block is that proposed by Friedman (1937). In this test the treatments within each block are ranked from lowest to highest. Then the sum of ranks is obtained for each treatment. The statistic used to test the null hypothesis that the t samples (into which the treatment classification divides the data) come from the same population is

$$\chi^2 = \frac{12}{bt(t+1)} \sum_{i=1}^{t} R_{i.}^2 - 3b(t+1)$$

where b is the number of blocks, t is the number of treatments, $R_{i.}$ is the sum of the ranks for the ith treatment. This test statistic is distributed approximately as a χ^2 variate with $(t-1)$ degrees of freedom. This approximation is worst for small b or t. but is sufficiently good for most of the commonly used randomized complete block experiments.

Example 12.6   Apply Friedman's procedure to the data contained in Exercise 9.10 (p. 181) for the species *O. sativus*.

The ranks within each block are presented in the following table.

Blocks	|	|			Strains				
	|	1	2	3	4	5	6	7	8
A	|	4	7	8	2	5	6	3	1
B	|	6	2	8	1	4	7	3	5
C	|	1	2	5	4	7	6	8	3
D	|	4	2	6	1	7	8	5	3
$R_{i.}$	|	15	13	27	8	23	27	19	12

$$\chi^2 = \frac{12}{4 \times 8(8+1)} \times 2950 - 3 \times 4 \times (8+1)$$

$$= 14 \cdot 92.$$

This χ^2 with seven degrees of freedom is significant at the 5% level. This result is very similar to that obtained by using the analysis of variance and F test. The observed F is 3·53 which has a probability of being exceeded of approximately 2%.

12.11 Other Non-Parametric Procedures

Only a few of the many non-parametric procedures which have been developed are presented here, since this is not a text on non-parametric statistics. Even if it were it would have to be large to adequately cover this rapidly-developing field of statistics. The reason for writing this chapter is to indicate the possibilities of these methods. Those who now think that non-parametric statistics could be useful to them in their research are referred to Siegel (1956), Savage (1962) and Walsh (1962, 1965).

Exercises

12.1 Use Lord's test based on range to test the hypothesis that the sample of 10 yields of butterfat given in Exercise 7.3 (p. 125) is a random sample from a population with mean 40 lb. Using the same procedure, estimate the 95% confidence interval for the mean. $(t_w = \pm 0·230, \alpha = 5\%.)$

12.2 Use Lord's test to test the hypothesis that the sample of 16 peach yields in Exercise 7.4 (p. 125) is a random sample from a population with mean 8·0 tons/acre. Using the same procedure estimate the 99% confidence interval for the population mean. $(t_w = \pm 0·212. \alpha = 1\%.)$

12.3 Using the data in Exercise 7.8 (p. 133) test by Lord's procedure the hypothesis that the two population means are equal. $(t_w = \pm 0·552, \alpha = 5\%.)$

12.4 Apply Lord's procedure to the yields of dry matter, yields of nitrogen and nitrogen fixation given in Exercise 7.15 (p. 142). In each case test the equality of the two population means. $(t_w = \pm 0·333, \alpha = 1\%.)$

12.5 In an experiment on mice from litters of different sizes, three treatments were used and the number of deaths per litter before weaning was observed.

(a) The following table summarizes the results of that part of the experiment where the litter size was 8. The entries in the body of the table are numbers of litters. Of the 67 litters of size 8 which were given treatment A, 54 had 0 deaths, 9 had 1 death and 4 had 2 or more deaths.

Treatment	Deaths per litter		
	0	1	2 or more
A	54	9	4
B	87	21	9
C	45	18	12

Use the χ^2 test to test the hypothesis that the number of deaths per litter classification is independent of treatment.

(b) The following table summarizes the results of the experiment as far as they related to treatment A. The classification in this case is according to pre-treatment litter size and deaths per litter.

Litter size	Deaths per litter		
	0	1	2 or more
7	45	12	2
8	54	9	4
9	30	15	12
10	16	14	13

Use the χ^2 test to test the hypothesis that the number of deaths per litter classification is independent of litter size.

12.6 In a cross between a rust resistant and a susceptible variety of wheat, the following data were obtained. 831 F_3 families were studied and classified according to their rust reaction in the field.

Seedling reaction	Field reaction		
	Resistant	Segregating	Susceptible
Resistant	156	16	5
Segregating	45	428	12
Susceptible	4	3	162

Test the independence of seedling and field reaction by means of a χ^2 test.

12.7   Use the Kruskal–Wallis procedure and the data in Exercise 9.2 to test the hypothesis that the groups of lambs on the four different diets are samples from the same population.

12.8   Apply Friedman's procedure to the data contained in Exercise 9.10 (p. 181) for the species *O. compressus*.

12.9   For each of the two variables—yield at first harvest and total yield after six harvests—and using the data of Exercise 9.11 use Friedman's procedure to test the hypothesis of no differences among the four treatment population means.

REFERENCES

AUBLE, J. D. (1953). 'Extended Tables for the Mann–Whitney Statistic', *Bulletin of the Institute of Educational Research* (Indiana University) 1, 1–39.

BRADLEY, J. V. (1960). *Distribution-Free Statistical Tests*, Wadd Technical Report 60–661, Wright Air Development Division.

FRIEDMAN, M. (1937). 'The Use of Ranks to Avoid the Assumption of Normality Implicit in the Analysis of Variance', *J. Amer. Stat. Assoc.*, 32, 675–701.

KOLMOGOROV, A. (1933). 'Sulla determinazione empirica di una legge di distributione', *Giornale dell'Istituto Italiano degli Attuari*, 4, 1–11.

KRUSKAL, W. H. and WALLIS, W. A. (1952). 'Use of Ranks in One-criterion Variance Analysis', *J. Amer. Stat. Assoc.*, 47, 583–621.

LORD, E. (1947). 'The Use of Range in Place of Standard Deviation in the *t*-test', *Biometrika*, 34, 41–67.

MANN, H. B. and WHITNEY, D. R. (1947). 'On a Test of Whether One of Two Random Variables is Stochastically Larger than the Other', *Ann. Math. Statist.*, 18, 50–60.

MASSEY, F. J. (1951). 'The Kolmogorov–Smirnov Test for Goodness of Fit', *J. Amer. Stat. Assoc.*, **46**, 68–78.

MOORE, P. G. (1957). 'The Two-sample *t*-test Based on Range', *Biometrika*, **44**, 482–489.

SAVAGE, I. R. A. (1962). *Bibliography of Non-Parametric Statistics*. Harvard University Press, Cambridge, Mass.

SIEGEL, S. (1956). *Nonparametric Statistics for the Behavioral Sciences*. McGraw-Hill, New York.

SMIRNOV, N. V. (1948). 'Table for Estimating the Goodness of Fit of Empirical Distributions', *Ann. Math. Statist.*, **19**, 279–281.

WALSH, J. E. (1962, 1965). *Handbook of Nonparametric Statistics*, Vols. I & II. D. van Nostrand, Princeton, New Jersey.

COLLATERAL READING

SNEDECOR, G. W. and COCHRAN, W. G. (1967). *Statistical Methods* (6th edition). Iowa State University Press, Ames, Iowa. Chapter 5.

SOKAL, R. R. and ROHLF, F. J. (1969). *Biometry*, W. H. Freeman, San Francisco. Chapter 13.

STEEL, R. G. D. and TORRIE, J. H. (1960). *Principles and Procedures of Statistics*. McGraw-Hill, New York. Chapter 21.

WOOLF, C. M. (1968). *Principles of Biometry*. D. van Nostrand, Princeton, New Jersey. Chapter 20.

Answers to Exercises

Chapter 2

1. Continuous, discrete, discrete, discrete, continuous.

2. $a_1 + a_2 + \ldots + a_5$; $(X_1 - a) + (X_2 - a) + (X_3 - a)$;
$X_1^2 + X_2^2 + \ldots + X_n^2$.

3. $\sum_{i=1}^{n} f_i Y_i^2$; $\sum_{i=1}^{n} X_i Y_i$; $\sum_{i=1}^{m} (X_i + Y_i + Z_i)$.

5. $346 \cdot 6$; $647 \cdot 9$; $419774 \cdot 41$; $20988 \cdot 72$; $21160 \cdot 33$.

6. $138 \cdot 4$ (bushels/acre); $162 \cdot 2$ (bushels/acre); $84515 \cdot 85$ (bushels/acre)^2.

9. $f_1(X_1 - \mu)^2 + f_2(X_2 - \mu)^2 + \ldots + f_4(X_4 - \mu)$;
$(f_1 X_1 + f_2 X_2 + f_3 X_3 + f_4 X_4)^2$; $f_1 X_1^2 + f_2 X_2^2 + f_3 X_3^2 + f_4 X_4^2$.

10. $171 \cdot 61$ (bushels/acre)^2; $171 \cdot 6095$ (bushels/acre)^2; they differ in the second decimal place because $32 \cdot 4$ is not the exact mean. Method (b) is preferred over method (a) for calculating $\sum (Y_j - \bar{y}.)^2$.

11. $\bar{x} = -0 \cdot 0058$; $s^2 = 1 \cdot 70$.

13. $9 \cdot 8$ grams; $s = 4 \cdot 78$ grams.

14. Median $= 8 \cdot 0$; $\bar{x} = 7 \cdot 97$; $s = 1 \cdot 48$.

15. $\bar{x} = 22 \cdot 85$ grams; $12 \cdot 75$ grams.

Chapter 3

1. $1/3$.　　2. $1/9$.　　3. $0 \cdot 3$.

4. $(\frac{1}{2})^{10}$; $0 \cdot 04$; $(\frac{1}{2})^{10}$; $(\frac{1}{2})^9$.

5. $(\frac{1}{6})^{24}$; $(\frac{1}{6})^{24}$.

6. $56 \cdot 97$; $91 \cdot 50$; $3248 \cdot 71$; $8375 \cdot 02$; $5215 \cdot 00$; covariance not zero, not independent.

7. $\sigma_u^2 + \sigma_v^2$; $\sigma_u^2 + \sigma_v^2$; $4\sigma_u^2 + 9\sigma_v^2$.

8. 20 (bushels/acre); $3 \cdot 36$ (bushels/acre)^2.

9. 5 (bushels/acre); 21 (bushels/acre)^2.

Chapter 4

1. 0·323.   **2.** 89·6.   **3.** 40; 4·90.
4. $\sqrt{40}$.   **5.** 0·048; 0·044.   **6.** 1·09.

Chapter 5

2. (a) 0·9772; 0·8185; 0·5; 0·0228.
  (b) 0·025; 0·8163.
3. (a) $D_1 = 4·12$ cm; probability of 97·5% that plant heights will increase by more than 4·12 cm; $D_2 = 4·12$ cm; $D_3 = 15·88$ cm; 95% probability that increase in plant height will be between 4·12 cm and 15·88 cm.
  (b) 16·65; 13·60.
4. 37·566; 3·841; 26·873; 16·928.
5. 0·05; 0·01; 0·98.
6. 13·442; 0·05; 0·94.
7. 4.

Chapter 6

1. 29·671 to 30·329; hypothesis rejected at 10% level.
2. 12·647 to 13·353; 0·26.
3. 0·05.
4. (a) σ is population standard deviation; s, the square root of the sample variance, is an estimator of σ.
  (b) 12·67; 11·36 to 13·98.
5. 0·00187.
6. Sum of the $\chi^2 = 44·4$ has 25 d.f.
7. Sum of the $\chi^2 = 29·6$ has 20 d.f.
8. $z_{obs} = 5·6$ is significant at 1% level.
9. $z_{obs} = 1·8$ is not significant at 1% level.
10. 0·2% to 9·8%; significant at 10% level.

Chapter 7

1. 2·228; −2·681; 1·725; 1·055.
2. 0·845; 0·475.
3. 44 pounds; 3·58 pounds; non-significant; 35·90 to 52·10 pounds.

4. 0·726 ton/acre; significant; 3·861 to 8·139 tons/acre.
5. 12·722 to 13·278.
6. Hypothesis is not rejected at 5% level; 39·16 to 44·84.
7. Hypothesis not rejected at 1% level.
8. Hypothesis rejected at 5% level ($t = 2·611$).

9. $\hat{\sigma}^2 = \sum\limits_{i=1}^{t} \sum\limits_{j=1}^{n_i} (X_{ij} - \bar{x}_{i.})^2 / (\sum n_i - t)$.

10. $t = 2·17$ has 10 d.f. and is significant at 5% level (one-tailed test).
11. Non-significant at 5% level (unpaired test); paired test and significant at 5% level; $-0·40$ to 4·40 and 0·72 to 3·28.
12. 0·01 (this probability is much greater).
13. Difference is highly significant.
14. 1·19; 1·69; hypothesis rejected at 5% level; 0·20 to 7·35.
15. H_0 is rejected at the 5% level for all three variables.

Chapter 8

1. 3·58.   **2.** 12.   **3.** $t^2 = F$.
5. Neither F is significant at 5% level.
6. For both variables, the F test is non-significant
7. $H_0: \sigma_1^2 = \sigma_2^2$; $H_1: \sigma_1^2 < \sigma_2^2$. H_0 rejected for both variables.
8. Difference is large enough to conclude A is better than B. ($F = 7.6$).

Chapter 9

1. Total S.S. $= 504$; Between S.S. $= 406$; $F = 24$ (d.f. 4,24).
2. $F = 7·5**$ (d.f. 3,13).
3. $F = 6·82$; $t = 2·61$; $6·82 = 2·61^2$.
4. (a) Yield of 2nd treatment in 3rd block; yield of 4th treatment in ith block; total yield for 3rd block; total yield of 2nd treatment; total yield for experiment.

   (b) Y_{34} ; Y_{i3} ; Y_{4j} ; $\sum\limits_{i=1}^{b} Y_{i1} = Y_{.1}$; $Y_{..}$

6. $\begin{bmatrix} 2 & 3 & 4 & 5 \\ 12 & 13 & 14 & 15 \end{bmatrix}$.

7. $F = 9·7$ (highly significant); l.s.d. $= 3·08$ (5%); 4·25 (1%); $A–E$ significant at 5%; $C–D$ significant at 1%.

8. $F = 2.21$, not significant.

9. $F = 13.6$, significant at 1% level; $F = t^2$.

10. $F = 3.53$ (*O. sativus*), significant; $F = 1.01$ (*O. compressus*), not significant; l.s.d. $= 467.4$.

11. $F = 30.2^{**}$ (first harvest); $F = 15.8^{**}$ (total).

Chapter 10

1. (a) $Y = 4.00 + 0.893X$, (b) $Y = -3.692 + 0.615X$,
(c) $Y = 4.371 - 0.686X$; No, (a) and (b) are positive,
(c) is negative; (a) and (b) are highly significant, (c) is significant.

2. $Y - 18.25 = 0.043(X - 225)$; $t = 1.133$, n.s.

3. $Y = 0.20 + 0.28X$; $F = 392.0^{**}$

4. 66.2 in; 157.8 lb; 0.92 in; 5.81 lb; 0.94; $F = 179^{**}$.

5. $Y = 11.83 + 0.851X$; $r = 0.90$; $F = 28.2^{**}$ (P).
$Y = -4.10 + 1.021X$; $r = 0.87$; $F = 37.9^{**}$ (N).

6. (a) $G = 41.93 - 0.450T$; $F = 908^{**}$; 0.0149; -0.416 to -0.484; 99%; -0.97 to -1.00.
(b) $L = -85.91 + 2.58T$; 0.177; $t = 2.37^*$; no; if hypothesized β in C.I., test not significant; $F = 213.6^{**}$; 0.964.

7. No, $z = 1.90$.

Chapter 11

1. $z = 0.12, \chi^2 = 0.015$, n.s.; $z = -0.51, \chi^2 = 0.26$, n.s.;
$z = -0.25, \chi^2 = 0.064$, n.s.; $z = -0.78, \chi^2 = 0.61$, n.s.;
$z = 0.62, \chi^2 = 0.39$, n.s.; $z = 0.59, \chi^2 = 0.35$, n.s.;
$z = -0.67, \chi^2 = 0.45$, n.s..

2. $\chi^2 = 13.53$, may not; $\chi^2 = 10.00$, may not; $\chi^2 = 1.23$, may; $\chi^2 = 7.47$, may; $\chi^2 = 3.73$, may.

3. Fits better $27:21:9:7$.

4. (a) $\chi^2 = 16.80^{**}$ ($v = 3$); (b) $\chi^2 = 3.75$ n.s. ($v = 2$).

5. 0.59; 1.04; 0.02; 0.42; 2.03; 0.05; 0.36; 1.82; 0.33; 0.54; χ^2 sum $= 7.19$; $v = 10$.

6. (a) $\chi^2 = 5.47, 0.51$; (b) $\chi^2 = 54.72^{**}, 5.11$.

7. $\chi^2 = 8.20^*$.

Chapter 12

1. $t_w = 0.108$ n.s.; 35·5 to 52·5.
2. $t_w = -0.213**$; 4·01 to 7·99.
3. $t_w = 0.581*$.
4. $t_w = 0.641**$, $0.800**$, $0.839**$.
5. (a) $\chi^2 = 8.99$, d.f. $= 4$, n.s.; (b) $\chi^2 = 33.5**$, d.f. $= 6$.
6. $\chi^2 = 1171.6**$, d.f. $= 4$.
7. $\chi^2 = 18.514**$, d.f. $= 3$.
8. $\chi^2 = 6.33$, d.f. $= 7$.
9. $\chi^2 = 25.2**$, $23.9*$, d.f. $= 3$.

Statistical Tables I–IV

TABLE I   *Probability of a random value of $z = (X - \mu)/\sigma$ lying between zero and the values tabulated in the margins*

Probability $= A = \int_0^z \phi(z)\, dz$

	0·00	0·01	0·02	0·03	0·04	0·05	0·06	0·07	0·08	0·09
0·0	0·0000	0·0040	0·0080	0·0120	0·0159	0·0199	0·0239	0·0279	0·0319	0·0359
0·1	0·0398	0·0438	0·0478	0·0517	0·0557	0·0596	0·0636	0·0675	0·0714	0·0753
0·2	0·0793	0·0832	0·0871	0·0910	0·0948	0·0987	0·1026	0·1064	0·1103	0·1141
0·3	0·1179	0·1217	0·1255	0·1293	0·1331	0·1368	0·1406	0·1443	0·1480	0·1517
0·4	0·1554	0·1591	0·1628	0·1664	0·1700	0·1736	0·1772	0·1808	0·1844	0·1879
0·5	0·1915	0·1950	0·1985	0·2019	0·2054	0·2088	0·2123	0·2157	0·2190	0·2224
0·6	0·2257	0·2291	0·2324	0·2357	0·2389	0·2422	0·2454	0·2486	0·2518	0·2549
0·7	0·2580	0·2611	0·2642	0·2673	0·2704	0·2734	0·2764	0·2794	0·2823	0·2852
0·8	0·2881	0·2910	0·2939	0·2967	0·2995	0·3023	0·3051	0·3078	0·3106	0·3133
0·9	0·3159	0·3186	0·3212	0·3238	0·3264	0·3289	0·3315	0·3340	0·3365	0·3389
1·0	0·3413	0·3438	0·3461	0·3485	0·3508	0·3531	0·3554	0·3577	0·3599	0·3621
1·1	0·3643	0·3665	0·3686	0·3708	0·3729	0·3749	0·3770	0·3790	0·3810	0·3830
1·2	0·3849	0·3869	0·3888	0·3907	0·3925	0·3944	0·3962	0·3980	0·3997	0·4015
1·3	0·4032	0·4049	0·4066	0·4082	0·4099	0·4115	0·4131	0·4147	0·4162	0·4177
1·4	0·4192	0·4207	0·4222	0·4236	0·4251	0·4265	0·4279	0·4292	0·4306	0·4319
1·5	0·4332	0·4345	0·4357	0·4370	0·4382	0·4394	0·4406	0·4418	0·4430	0·4441
1·6	0·4452	0·4463	0·4474	0·4485	0·4495	0·4505	0·4515	0·4525	0·4535	0·4545
1·7	0·4554	0·4563	0·4573	0·4582	0·4591	0·4599	0·4608	0·4616	0·4625	0·4633
1·8	0·4641	0·4649	0·4656	0·4664	0·4671	0·4678	0·4686	0·4693	0·4699	0·4706
1·9	0·4713	0·4719	0·4726	0·4732	0·4738	0·4744	0·4750	0·4756	0·4762	0·4767
2·0	0·4772	0·4778	0·4783	0·4788	0·4793	0·4798	0·4803	0·4808	0·4812	0·4817
2·1	0·4821	0·4826	0·4830	0·4834	0·4838	0·4842	0·4846	0·4850	0·4854	0·4857
2·2	0·4861	0·4865	0·4868	0·4871	0·4875	0·4878	0·4881	0·4884	0·4887	0·4890
2·3	0·4893	0·4896	0·4898	0·4901	0·4904	0·4906	0·4909	0·4911	0·4913	0·4916
2·4	0·4918	0·4920	0·4922	0·4925	0·4927	0·4929	0·4931	0·4932	0·4934	0·4936
2·5	0·4938	0·4940	0·4941	0·4943	0·4945	0·4946	0·4948	0·4949	0·4951	0·4952
2·6	0·4953	0·4955	0·4956	0·4957	0·4959	0·4960	0·4961	0·4962	0·4963	0·4964
2·7	0·4965	0·4966	0·4967	0·4968	0·4969	0·4970	0·4971	0·4972	0·4973	0·4974
2·8	0·4974	0·4975	0·4976	0·4977	0·4977	0·4978	0·4979	0·4980	0·4980	0·4981
2·9	0·4981	0·4982	0·4983	0·4983	0·4984	0·4984	0·4985	0·4985	0·4986	0·4986
3·0	0·4986	0·4987	0·4987	0·4988	0·4988	0·4989	0·4989	0·4989	0·4990	0·4990
3·1	0·4990	0·4991	0·4991	0·4991	0·4992	0·4992	0·4992	0·4992	0·4993	0·4993

TABLE II   Percentage points χ^2_α of the χ^2 distribution

ν	·99	·98	·95	·90	·80	·70	·50	·30	·20	·10	·05	·02	·01	·001
1	0·0³157ʰ	0·0³628	0·00393	0·0158	0·0642	0·148	0·455	1·074	1·642	2·706	3·841	5·412	6·635	10·827
2	0·0201	0·0404	0·103	0·211	0·446	0·713	1·386	2·408	3·219	4·605	5·991	7·824	9·210	13·815
3	0·115	0·185	0·352	0·584	1·005	1·424	2·366	3·665	4·642	6·251	7·815	9·837	11·345	16·266
4	0·297	0·429	0·711	1·064	1·649	2·195	3·357	4·878	5·989	7·779	9·488	11·668	13·277	18·467
5	0·554	0·752	1·145	1·610	2·343	3·000	4·351	6·064	7·289	9·236	11·070	13·388	15·086	20·515
6	0·872	1·134	1·635	2·204	3·070	3·828	5·348	7·231	8·558	10·645	12·592	15·033	16·812	22·457
7	1·239	1·564	2·167	2·833	3·822	4·671	6·346	8·383	9·803	12·017	14·067	16·622	18·475	24·322
8	1·646	2·032	2·733	3·490	4·594	5·527	7·344	9·524	11·030	13·362	15·507	18·168	20·090	26·125
9	2·088	2·532	3·325	4·168	5·380	6·393	8·343	10·656	12·242	14·684	16·919	19·679	21·666	27·877
10	2·558	3·059	3·940	4·865	6·179	7·267	9·342	11·781	13·442	15·987	18·307	21·161	23·209	29·588
11	3·053	3·609	4·575	5·578	6·989	8·148	10·341	12·899	14·631	17·275	19·675	22·618	24·725	31·264
12	3·571	4·178	5·226	6·304	7·807	9·034	11·340	14·011	15·812	18·549	21·026	24·054	26·217	32·909
13	4·107	4·765	5·892	7·042	8·634	9·926	12·340	15·119	16·985	19·812	22·362	25·472	27·688	34·528
14	4·660	5·368	6·571	7·790	9·467	10·821	13·339	16·222	18·151	21·064	23·685	26·873	29·141	36·123
15	5·229	5·985	7·261	8·547	10·307	11·721	14·339	17·322	19·311	22·307	24·996	28·259	30·578	37·697
16	5·812	6·614	7·962	9·312	11·152	12·624	15·338	18·418	20·465	23·542	26·296	29·633	32·000	39·252
17	6·408	7·255	8·672	10·085	12·002	13·531	16·338	19·511	21·615	24·769	27·587	30·995	33·409	40·790
18	7·015	7·906	9·390	10·865	12·857	14·440	17·338	20·601	22·760	25·989	28·869	32·346	34·805	42·312
19	7·633	8·567	10·117	11·651	13·716	15·352	18·338	21·689	23·900	27·204	30·144	33·687	36·191	43·820
20	8·260	9·237	10·851	12·443	14·578	16·266	19·337	22·775	25·038	28·412	31·410	35·020	37·566	45·315
21	8·897	9·915	11·591	13·240	15·445	17·182	20·337	23·858	26·171	29·615	32·671	36·343	38·932	46·797
22	9·542	10·600	12·338	14·041	16·314	18·101	21·337	24·939	27·301	30·813	33·924	37·659	40·289	48·268
23	10·196	11·293	13·091	14·848	17·187	19·021	22·337	26·018	28·429	32·007	35·172	38·968	41·638	49·728
24	10·856	11·992	13·848	15·659	18·062	19·943	23·337	27·096	29·553	33·196	36·415	40·270	42·980	51·179
25	11·524	12·697	14·611	16·473	18·940	20·867	24·337	28·172	30·675	34·382	37·652	41·566	44·314	52·620

ν														
26	12·198	13·409	15·379	17·292	19·820	21·792	25·336	29·246	31·795	35·563	38·885	42·856	45·642	54·052
27	12·879	14·125	16·151	18·114	20·703	22·719	26·336	30·319	32·912	36·741	40·113	44·140	46·963	55·476
28	13·565	14·847	16·928	18·939	21·588	23·647	27·336	31·391	34·027	37·916	41·337	45·419	48·278	56·893
29	14·256	15·574	17·708	19·768	22·475	24·577	28·336	32·461	35·139	39·087	42·557	46·693	49·588	58·302
30	14·953	16·306	18·493	20·599	23·364	25·508	29·336	33·530	36·250	40·256	43·773	47·962	50·892	59·703
32	16·362	17·783	20·072	22·271	25·148	27·373	31·336	35·665	38·466	42·585	46·194	50·487	53·486	62·487
34	17·789	19·275	21·664	23·952	26·938	29·242	33·336	37·795	40·676	44·903	48·602	52·995	56·061	65·247
36	19·233	20·783	23·269	25·643	28·735	31·115	35·336	39·922	42·879	47·212	50·999	55·489	58·619	67·985
38	20·691	22·304	24·884	27·343	30·537	32·992	37·335	42·045	45·076	49·513	53·384	57·969	61·162	70·703
40	22·164	23·838	26·509	29·051	32·345	34·872	39·335	44·165	47·269	51·805	55·759	60·436	63·691	73·402
42	23·650	25·383	28·144	30·765	34·157	36·755	41·335	46·282	49·456	54·090	58·124	62·892	66·206	76·084
44	25·148	26·939	29·787	32·487	35·974	38·641	43·335	48·396	51·639	56·369	60·481	65·337	68·710	78·750
46	26·657	28·504	31·439	34·215	37·795	40·529	45·335	50·507	53·818	58·641	62·830	67·771	71·201	81·400
48	28·177	30·080	33·098	35·949	39·621	42·420	47·335	52·616	55·993	60·907	65·171	70·197	73·683	84·037
50	29·707	31·664	34·764	37·689	41·449	44·313	49·335	54·723	58·164	63·167	67·505	72·613	76·154	86·661
52	31·246	33·256	36·437	39·433	43·281	46·209	51·335	56·827	60·332	65·422	69·832	75·021	78·616	89·272
54	32·793	34·856	38·116	41·183	45·117	48·106	53·335	58·930	62·496	67·673	72·153	77·422	81·069	91·872
56	34·350	36·464	39·801	42·937	46·955	50·005	55·335	61·031	64·658	69·919	74·468	79·815	83·513	94·461
58	35·913	38·078	41·492	44·696	48·797	51·906	57·335	63·129	66·816	72·160	76·778	82·201	85·950	97·039
60	37·485	39·699	43·188	46·459	50·641	53·809	59·335	65·227	68·972	74·397	79·082	84·580	88·379	99·607
62	39·063	41·327	44·889	48·226	52·487	55·714	61·335	67·322	71·125	76·630	81·381	86·953	90·802	102·166
64	40·649	42·960	46·595	49·996	54·336	57·620	63·335	69·416	73·276	78·860	83·675	89·320	93·217	104·716
66	42·240	44·599	48·305	51·770	56·188	59·527	65·335	71·508	75·424	81·085	85·965	91·681	95·626	107·258
68	43·838	46·244	50·020	53·548	58·042	61·436	67·335	73·600	77·571	83·308	88·250	94·037	98·028	109·791
70	45·442	47·893	51·739	55·329	59·898	63·346	69·334	75·689	79·715	85·527	90·531	96·388	100·425	112·317

For odd values of ν between 30 and 70 the mean of the tabular values for $\nu - 1$ and $\nu + 1$ may be taken. For larger values of ν, the expression $\sqrt{2\chi^2} - \sqrt{2\nu - 1}$ may be used as a normal deviate with unit variance, remembering that the probability for χ^2 corresponds with that of a single tail of the normal curve.

Table II is taken from Table IV of Fisher and Yates: Statistical Tables for Biological, Agricultural and Medical Research, published by Oliver & Boyd Ltd., Edinburgh, and by permission of the authors and publishers.

R

TABLE III   Percentage points $t_{\frac{1}{2}\alpha}$ of the t distribution

v	Probability												
	0·9	0·8	0·7	0·6	0·5	0·4	0·3	0·2	0·1	0·05	0·02	0·01	0·001
1	0·158	0·325	0·510	0·727	1·000	1·376	1·963	3·078	6·314	12·706	31·821	63·657	636·619
2	0·142	0·289	0·445	0·617	0·816	1·061	1·386	1·886	2·920	4·303	6·965	9·925	31·598
3	0·137	0·277	0·424	0·584	0·765	0·978	1·250	1·638	2·353	3·182	4·541	5·841	12·941
4	0·134	0·271	0·414	0·569	0·741	0·941	1·190	1·533	2·132	2·776	3·747	4·604	8·610
5	0·132	0·267	0·408	0·559	0·727	0·920	1·156	1·476	2·015	2·571	3·365	4·032	6·859
6	0·131	0·265	0·404	0·553	0·718	0·906	1·134	1·440	1·943	2·447	3·143	3·707	5·959
7	0·130	0·263	0·402	0·549	0·711	0·896	1·119	1·415	1·895	2·365	2·998	3·499	5·405
8	0·130	0·262	0·399	0·546	0·706	0·889	1·108	1·397	1·860	2·306	2·896	3·355	5·041
9	0·129	0·261	0·398	0·543	0·703	0·883	1·100	1·383	1·833	2·262	2·821	3·250	4·781
10	0·129	0·260	0·397	0·542	0·700	0·879	1·093	1·372	1·812	2·228	2·764	3·169	4·587
11	0·129	0·260	0·396	0·540	0·697	0·876	1·088	1·363	1·796	2·201	2·718	3·106	4·437
12	0·128	0·259	0·395	0·539	0·695	0·873	1·083	1·356	1·782	2·179	2·681	3·055	4·318
13	0·128	0·259	0·394	0·538	0·694	0·870	1·079	1·350	1·771	2·160	2·650	3·012	4·221
14	0·128	0·258	0·393	0·537	0·692	0·868	1·076	1·345	1·761	2·145	2·624	2·977	4·140
15	0·128	0·258	0·393	0·536	0·691	0·866	1·074	1·341	1·753	2·131	2·602	2·947	4·073

16	0·128	0·258	0·392	0·535	0·690	0·865	1·071	1·337	1·746	2·120	2·583	2·921	4·015
17	0·128	0·257	0·392	0·534	0·689	0·863	1·069	1·333	1·740	2·110	2·567	2·898	3·965
18	0·127	0·257	0·392	0·534	0·688	0·862	1·067	1·330	1·734	2·101	2·552	2·878	3·922
19	0·127	0·257	0·391	0·533	0·688	0·861	1·066	1·328	1·729	2·093	2·539	2·861	3·883
20	0·127	0·257	0·391	0·533	0·687	0·860	1·064	1·325	1·725	2·086	2·528	2·845	3·850
21	0·127	0·257	0·391	0·532	0·686	0·859	1·063	1·323	1·721	2·080	2·518	2·831	3·819
22	0·127	0·256	0·390	0·532	0·686	0·858	1·061	1·321	1·717	2·074	2·508	2·819	3·792
23	0·127	0·256	0·390	0·532	0·685	0·858	1·060	1·319	1·714	2·069	2·500	2·807	3·767
24	0·127	0·256	0·390	0·531	0·685	0·857	1·059	1·318	1·711	2·064	2·492	2·797	3·745
25	0·127	0·256	0·390	0·531	0·684	0·856	1·058	1·316	1·708	2·060	2·485	2·787	3·725
26	0·127	0·256	0·390	0·531	0·684	0·856	1·058	1·315	1·706	2·056	2·479	2·779	3·707
27	0·127	0·256	0·389	0·531	0·684	0·855	1·057	1·314	1·703	2·052	2·473	2·771	3·690
28	0·127	0·256	0·389	0·530	0·683	0·855	1·056	1·313	1·701	2·048	2·467	2·763	3·674
29	0·127	0·256	0·389	0·530	0·683	0·854	1·055	1·311	1·699	2·045	2·462	2·756	3·659
30	0·127	0·256	0·389	0·530	0·683	0·854	1·055	1·310	1·697	2·042	2·457	2·750	3·646
40	0·126	0·255	0·388	0·529	0·681	0·851	1·050	1·303	1·684	2·021	2·423	2·704	3·551
60	0·126	0·254	0·387	0·527	0·679	0·848	1·046	1·296	1·671	2·000	2·390	2·660	3·460
120	0·126	0·254	0·386	0·526	0·677	0·845	1·041	1·289	1·658	1·980	2·358	2·617	3·373
∞	0·126	0·253	0·385	0·524	0·674	0·842	1·036	1·282	1·645	1·960	2·326	2·576	3·291

Table III is taken from Table III of Fisher and Yates. *Statistical Tables for Biological, Agricultural and Medical Research*, published by Oliver & Boyd Ltd., Edinburgh, and by permission of the authors and publishers.

TABLE IV   *Percentage points F_α of the F distribution*

Denominator df	Probability of a larger F	Numerator df								
		1	2	3	4	5	6	7	8	9
1	0·100	39·86	49·50	53·59	55·83	57·24	58·20	58·91	59·44	59·86
	0·050	161·4	199·5	215·7	224·6	230·2	234·0	236·8	238·9	240·5
	0·025	647·8	799·5	864·2	899·6	921·8	937·1	948·2	956·7	963·3
	0·010	4052	4999·5	5403	5625	5764	5859	5928	5982	6022
	0·005	16211	20000	21615	22500	23056	23437	23715	23925	24091
2	0·100	8·53	9·00	9·16	9·24	9·29	9·33	9·35	9·37	9·38
	0·050	18·51	19·00	19·16	19·25	19·30	19·33	19·35	19·37	19·38
	0·025	38·51	39·00	39·17	39·25	39·30	39·33	39·36	39·37	39·39
	0·010	98·50	99·00	99·17	99·25	99·30	99·33	99·36	99·37	99·39
	0·005	198·5	199·0	199·2	199·2	199·3	199·3	199·4	199·4	199·4
3	0·100	5·54	5·46	5·39	5·34	5·31	5·28	5·27	5·25	5·24
	0·050	10·13	9·55	9·28	9·12	9·01	8·94	8·89	8·85	8·81
	0·025	17·44	16·04	15·44	15·10	14·88	14·73	14·62	14·54	14·47
	0·010	34·12	30·82	29·46	28·71	28·24	27·91	27·67	27·49	27·35
	0·005	55·55	49·80	47·47	46·19	45·39	44·84	44·43	44·13	43·88
4	0·100	4·54	4·32	4·19	4·11	4·05	4·01	3·98	3·95	3·94
	0·050	7·71	6·94	6·59	6·39	6·26	6·16	6·09	6·04	6·00
	0·025	12·22	10·65	9·98	9·60	9·36	9·20	9·07	8·98	8·90
	0·010	21·20	18·00	16·69	15·98	15·52	15·21	14·98	14·80	14·66
	0·005	31·33	26·28	24·26	23·15	22·46	21·97	21·62	21·35	21·14
5	0·100	4·06	3·78	3·62	3·52	3·45	3·40	3·37	3·34	3·32
	0·050	6·61	5·79	5·41	5·19	5·05	4·95	4·88	4·82	4·77
	0·025	10·01	8·43	7·76	7·39	7·15	6·98	6·85	6·76	6·68
	0·010	16·26	13·27	12·06	11·39	10·97	10·67	10·46	10·29	10·16
	0·005	22·78	18·31	16·53	15·56	14·94	14·51	14·20	13·96	13·77
6	0·100	3·78	3·46	3·29	3·18	3·11	3·05	3·01	2·98	2·96
	0·050	5·99	5·14	4·76	4·53	4·39	4·28	4·21	4·15	4·10
	0·025	8·81	7·26	6·60	6·23	5·99	5·82	5·70	5·60	5·52
	0·010	13·75	10·92	9·78	9·15	8·75	8·47	8·26	8·10	7·98
	0·005	18·63	14·54	12·92	12·03	11·46	11·07	10·79	10·57	10·39
7	0·100	3·59	3·26	3·07	2·96	2·88	2·83	2·78	2·75	2·72
	0·050	5·59	4·74	4·35	4·12	3·97	3·87	3·79	3·73	3·68
	0·025	8·07	6·54	5·89	5·52	5·29	5·12	4·99	4·90	4·82
	0·010	12·25	9·55	8·45	7·85	7·46	7·19	6·99	6·84	6·72
	0·005	16·24	12·40	10·88	10·05	9·52	9·16	8·89	8·68	8·51
8	0·100	3·46	3·11	2·92	2·81	2·73	2·67	2·62	2·59	2·56
	0·050	5·32	4·46	4·07	3·84	3·69	3·58	3·50	3·44	3·39
	0·025	7·57	6·06	5·42	5·05	4·82	4·65	4·53	4·43	4·36
	0·010	11·26	8·65	7·59	7·01	6·63	6·37	6·18	6·03	5·91
	0·005	14·69	11·04	9·60	8·81	8·30	7·95	7·69	7·50	7·34
9	0·100	3·36	3·01	2·81	2·69	2·61	2·55	2·51	2·47	2·44
	0·050	5·12	4·26	3·86	3·63	3·48	3·37	3·29	3·23	3·18
	0·025	7·21	5·71	5·08	4·72	4·48	4·32	4·20	4·10	4·03
	0·010	10·56	8·02	6·99	6·42	6·06	5·80	5·61	5·47	5·35
	0·005	13·61	10·11	8·72	7·96	7·47	7·13	6·88	6·69	6·54

Numerator df											
10	12	15	20	24	30	40	60	120	∞	P	df
60·19	60·71	61·22	61·74	62·00	62·26	62·53	62·79	63·06	63·33	0·100	1
241·9	243·9	245·9	248·0	249·1	250·1	251·1	252·2	253·3	254·3	0·050	
968·6	976·7	984·9	993·1	997·2	1001	1006	1010	1014	1018	0·025	
6056	6106	6157	6209	6235	6261	6287	6313	6339	6366	0·010	
24224	24426	24630	24836	24940	25044	25148	25253	25359	25465	0·005	
9·39	9·41	9·42	9·44	9·45	9·46	9·47	9·47	9·48	9·49	0·100	2
19·40	19·41	19·43	19·45	19·45	19·46	19·47	19·48	19·49	19·50	0·050	
39·40	39·41	39·43	39·45	39·46	39·46	39·47	39·48	39·49	39·50	0·025	
99·40	99·42	99·43	99·45	99·46	99·47	99·47	99·48	99·49	99·50	0·010	
199·4	199·4	199·4	199·4	199·5	199·5	199·5	199·5	199·5	199·5	0·005	
5·23	5·22	5·20	5·18	5·18	5·17	5·16	5·15	5·14	5·13	0·100	3
8·79	8·74	8·70	8·66	8·64	8·62	8·59	8·57	8·55	8·53	0·050	
14·42	14·34	14·25	14·17	14·12	14·08	14·04	13·99	13·95	13·90	0·025	
27·23	27·05	26·87	26·69	26·60	26·50	26·41	26·32	26·22	26·13	0·010	
43·69	43·39	43·08	42·78	42·62	42·47	42·31	42·15	41·99	41·83	0·005	
3·92	3·90	3·87	3·84	3·83	3·82	3·80	3·79	3·78	3·76	0·100	4
5·96	5·91	5·86	5·80	5·77	5·75	5·72	5·69	5·66	5·63	0·050	
8·84	8·75	8·66	8·56	8·51	8·46	8·41	8·36	8·31	8·26	0·025	
14·55	14·37	14·20	14·02	13·93	13·84	13·75	13·65	13·56	13·46	0·010	
20·97	20·70	20·44	20·17	20·03	19·89	19·75	19·61	19·47	19·32	0·005	
3·30	3·27	3·24	3·21	3·19	3·17	3·16	3·14	3·12	3·10	0·100	5
4·74	4·68	4·62	4·56	4·53	4·50	4·46	4·43	4·40	4·36	0·050	
6·62	6·52	6·43	6·33	6·28	6·23	6·18	6·12	6·07	6·02	0·025	
10·05	9·89	9·72	9·55	9·47	9·38	9·29	9·20	9·11	9·02	0·010	
13·62	13·38	13·15	12·90	12·78	12·66	12·53	12·40	12·27	12·14	0·005	
2·94	2·90	2·87	2·84	2·82	2·80	2·78	2·76	2·74	2·72	0·100	6
4·06	4·00	3·94	3·87	3·84	3·81	3·77	3·74	3·70	3·67	0·050	
5·46	5·37	5·27	5·17	5·12	5·07	5·01	4·96	4·90	4·85	0·025	
7·87	7·72	7·56	7·40	7·31	7·23	7·14	7·06	6·97	6·88	0·010	
10·25	10·03	9·81	9·59	9·47	9·36	9·24	9·12	9·00	8·88	0·005	
2·70	2·67	2·63	2·59	2·58	2·56	2·54	2·51	2·49	2·47	0·100	7
3·64	3·57	3·51	3·44	3·41	3·38	3·34	3·30	3·27	3·23	0·050	
4·76	4·67	4·57	4·47	4·42	4·36	4·31	4·25	4·20	4·14	0·025	
6·62	6·47	6·31	6·16	6·07	5·99	5·91	5·82	5·74	5·65	0·010	
8·38	8·18	7·97	7·75	7·65	7·53	7·42	7·31	7·19	7·08	0·005	
2·54	2·50	2·46	2·42	2·40	2·38	2·36	2·34	2·32	2·29	0·100	8
3·35	3·28	3·22	3·15	3·12	3·08	3·04	3·01	2·97	2·93	0·050	
4·30	4·20	4·10	4·00	3·95	3·89	3·84	3·78	3·73	3·67	0·025	
5·81	5·67	5·52	5·36	5·28	5·20	5·12	5·03	4·95	4·86	0·010	
7·21	7·01	6·81	6·61	6·50	6·40	6·29	6·18	6·06	5·95	0·005	
2·42	2·38	2·34	2·30	2·28	2·25	2·23	2·21	2·18	2·16	0·100	9
3·14	3·07	3·01	2·94	2·90	2·86	2·83	2·79	2·75	2·71	0·050	
3·96	3·87	3·77	3·67	3·61	3·56	3·51	3·45	3·39	3·33	0·025	
5·26	5·11	4·96	4·81	4·73	4·65	4·57	4·48	4·40	4·31	0·010	
6·42	6·23	6·03	5·83	5·73	5·62	5·52	5·41	5·30	5·19	0·005	

Percentage points F_α of the F distribution (cont.)

Denominator df	Probability of a larger F	Numerator df								
		1	2	3	4	5	6	7	8	9
10	0·100	3·29	2·92	2·73	2·61	2·52	2·46	2·41	2·38	2·35
	0·500	4·96	4·10	3·71	3·48	3·33	3·22	3·14	3·07	3·02
	0·025	6·94	5·46	4·83	4·47	4·24	4·07	3·95	3·85	3·78
	0·010	10·04	7·56	6·55	5·99	5·64	5·39	5·20	5·06	4·94
	0·005	12·83	9·43	8·08	7·34	6·87	6·54	6·30	6·12	5·97
11	0·100	3·23	2·86	2·66	2·54	2·45	2·39	2·34	2·30	2·27
	0·050	4·84	3·98	3·59	3·36	3·20	3·09	3·01	2·95	2·90
	0·025	6·72	5·26	4·63	4·28	4·04	3·88	3·76	3·66	3·59
	0·010	9·65	7·21	6·22	5·67	5·32	5·07	4·89	4·74	4·63
	0·005	12·23	8·91	7·60	6·88	6·42	6·10	5·86	5·68	5·54
12	0·100	3·18	2·81	2·61	2·48	2·39	2·33	2·28	2·24	2·21
	0·050	4·75	3·89	3·49	3·26	3·11	3·00	2·91	2·85	2·80
	0·025	6·55	5·10	4·47	4·12	3·89	3·73	3·61	3·51	3·44
	0·010	9·33	6·93	5·95	5·41	5·06	4·82	4·64	4·50	4·39
	0·005	11·75	8·51	7·23	6·52	6·07	5·76	5·52	5·35	5·20
13	0·100	3·14	2·76	2·56	2·43	2·35	2·28	2·23	2·20	2·16
	0·050	4·67	3·81	3·41	3·18	3·03	2·92	2·83	2·77	2·71
	0·025	6·41	4·97	4·35	4·00	3·77	3·60	3·48	3·39	3·31
	0·010	9·07	6·70	5·74	5·21	4·86	4·62	4·44	4·30	4·19
	0·005	11·37	8·19	6·93	6·23	5·79	5·48	5·25	5·08	4·94
14	0·100	3·10	2·73	2·52	2·39	2·31	2·24	2·19	2·15	2·12
	0·050	4·60	3·74	3·34	3·11	2·96	2·85	2·76	2·70	2·65
	0·025	6·30	4·86	4·24	3·89	3·66	3·50	3·38	3·29	3·21
	0·010	8·86	6·51	5·56	5·04	4·69	4·46	4·28	4·14	4·03
	0·005	11·06	7·92	6·68	6·00	5·56	5·26	5·03	4·86	4·72
15	0·100	3·07	2·70	2·49	2·36	2·27	2·21	2·16	2·12	2·09
	0·050	4·54	3·68	3·29	3·06	2·90	2·79	2·71	2·64	2·59
	0·025	6·20	4·77	4·15	3·80	3·58	3·41	3·29	3·20	3·12
	0·010	8·68	6·36	5·42	4·89	4·56	4·32	4·14	4·00	3·89
	0·005	10·80	7·70	6·48	5·80	5·37	5·07	4·85	4·67	4·54
16	0·100	3·05	2·67	2·46	2·33	2·24	2·18	2·13	2·09	2·06
	0·050	4·49	3·63	3·24	3·01	2·85	2·74	2·66	2·59	2·54
	0·025	6·12	4·69	4·08	3·73	3·50	3·34	3·22	3·12	3·05
	0·010	8·53	6·23	5·29	4·77	4·44	4·20	4·03	3·89	3·78
	0·005	10·58	7·51	6·30	5·64	5·21	4·91	4·69	4·52	4·38
17	0·100	3·03	2·64	2·44	2·31	2·22	2·15	2·10	2·06	2·03
	0·050	4·45	3·59	3·20	2·96	2·81	2·70	2·61	2·55	2·49
	0·025	6·04	4·62	4·01	3·66	3·44	3·28	3·16	3·06	2·98
	0·010	8·40	6·11	5·18	4·67	4·34	4·10	3·93	3·79	3·68
	0·005	10·38	7·35	6·16	5·50	5·07	4·78	4·56	4·39	4·25
18	0·100	3·01	2·62	2·42	2·29	2·20	2·13	2·08	2·04	2·00
	0·050	4·41	3·55	3·16	2·93	2·77	2·66	2·58	2·51	2·46
	0·025	5·98	4·56	3·95	3·61	3·38	3·22	3·10	3·01	2·93
	0·010	8·29	6·01	5·09	4·58	4·25	4·01	3·84	3·71	3·60
	0·005	10·22	7·21	6·03	5·37	4·96	4·66	4·44	4·28	4·14

| | | | | Numerator *df* | | | | | | | |
10	12	15	20	24	30	40	60	120	∞	P	df
2·32	2·28	2·24	2·20	2·18	2·16	2·13	2·11	2·08	2·06	0·100	10
2·98	2·91	2·85	2·77	2·74	2·70	2·66	2·62	2·58	2·54	0·050	
3·72	3·62	3·52	3·42	3·37	3·31	3·26	3·20	3·14	3·08	0·025	
4·85	4·71	4·56	4·41	4·33	4·25	4·17	4·08	4·00	3·91	0·010	
5·85	5·66	5·47	5·27	5·17	5·07	4·97	4·86	4·75	4·64	0·005	
2·25	2·21	2·17	2·12	2·10	2·08	2·05	2·03	2·00	1·97	0·100	11
2·85	2·79	2·72	2·65	2·61	2·57	2·53	2·49	2·45	2·40	0·050	
3·53	3·43	3·33	3·23	3·17	3·12	3·06	3·00	2·94	2·88	0·025	
4·54	4·40	4·25	4·10	4·02	3·94	3·86	3·78	3·69	3·60	0·010	
5·42	5·24	5·05	4·86	4·76	4·65	4·55	4·44	4·34	4·23	0·005	
2·19	2·15	2·10	2·06	2·04	2·01	1·99	1·96	1·93	1·90	0·100	12
2·75	2·69	2·62	2·54	2·51	2·47	2·43	2·38	2·34	2·30	0·050	
3·37	3·28	3·18	3·07	3·02	2·96	2·91	2·85	2·79	2·72	0·025	
4·30	4·16	4·01	3·86	3·78	3·70	3·62	3·54	3·45	3·36	0·010	
5·09	4·91	4·72	4·53	4·43	4·33	4·23	4·12	4·01	3·90	0·005	
2·14	2·10	2·05	2·01	1·98	1·96	1·93	1·90	1·88	1·85	0·100	13
2·67	2·60	2·53	2·46	2·42	2·38	2·34	2·30	2·25	2·21	0·050	
3·25	3·15	3·05	2·95	2·89	2·84	2·78	2·72	2·66	2·60	0·025	
4·10	3·96	3·82	3·66	3·59	3·51	3·43	3·34	3·25	3·17	0·010	
4·82	4·64	4·46	4·27	4·17	4·07	3·97	3·87	3·76	3·65	0·005	
2·10	2·05	2·01	1·96	1·94	1·91	1·89	1·86	1·83	1·80	0·100	14
2·60	2·53	2·46	2·39	2·35	2·31	2·27	2·22	2·18	2·13	0·050	
3·15	3·05	2·95	2·84	2·79	2·73	2·67	2·61	2·55	2·49	0·025	
3·94	3·80	3·66	3·51	3·43	3·35	3·27	3·18	3·09	3·00	0·010	
4·60	4·43	4·25	4·06	3·96	3·86	3·76	3·66	3·55	3·44	0·005	
2·06	2·02	1·97	1·92	1·90	1·87	1·85	1·82	1·79	1·76	0·100	15
2·54	2·48	2·40	2·33	2·29	2·25	2·20	2·16	2·11	2·07	0·050	
3·06	2·96	2·86	2·76	2·70	2·64	2·59	2·52	2·46	2·40	0·025	
3·80	3·67	3·52	3·37	3·29	3·21	3·13	3·05	2·96	2·87	0·010	
4·42	4·25	4·07	3·88	3·79	3·69	3·58	3·48	3·37	3·26	0·005	
2·03	1·99	1·94	1·89	1·87	1·84	1·81	1·78	1·75	1·72	0·100	16
2·49	2·42	2·35	2·28	2·24	2·19	2·15	2·11	2·06	2·01	0·050	
2·99	2·89	2·79	2·68	2·63	2·57	2·51	2·45	2·38	2·32	0·025	
3·69	3·55	3·41	3·26	3·18	3·10	3·02	2·93	2·84	2·75	0·010	
4·27	4·10	3·92	3·73	3·64	3·54	3·44	3·33	3·22	3·11	0·005	
2·00	1·96	1·91	1·86	1·84	1·81	1·78	1·75	1·72	1·69	0·100	17
2·45	2·38	2·31	2·23	2·19	2·15	2·10	2·06	2·01	1·96	0·050	
2·92	2·82	2·72	2·62	2·56	2·50	2·44	2·38	2·32	2·25	0·025	
3·59	3·46	3·31	3·16	3·08	3·00	2·92	2·83	2·75	2·65	0·010	
4·14	3·97	3·79	3·61	3·51	3·41	3·31	3·21	3·10	2·98	0·005	
1·98	1·93	1·89	1·84	1·81	1·78	1·75	1·72	1·69	1·66	0·100	18
2·41	2·34	2·27	2·19	2·15	2·11	2·06	2·02	1·97	1·92	0·050	
2·87	2·77	2·67	2·56	2·50	2·44	2·38	2·32	2·26	2·19	0·025	
3·51	3·37	3·23	3·08	3·00	2·92	2·84	2·75	2·66	2·57	0·010	
4·03	3·86	3·68	3·50	3·40	3·30	3·20	3·10	2·99	2·87	0·005	

Percentage points F_α of the F distribution (cont.)

Denominator df	Probability of a larger F	Numerator df								
		1	2	3	4	5	6	7	8	9
19	0·100	2·99	2·61	2·40	2·27	2·18	2·11	2·06	2·02	1·98
	0·050	4·38	3·52	3·13	2·90	2·74	2·63	2·54	2·48	2·42
	0·025	5·92	4·51	3·90	3·56	3·33	3·17	3·05	2·96	2·88
	0·010	8·18	5·93	5·01	4·50	4·17	3·94	3·77	3·63	3·52
	0·005	10·07	7·09	5·92	5·27	4·85	4·56	4·34	4·18	4·04
20	0·100	2·97	2·59	2·38	2·25	2·16	2·09	2·04	2·00	1·96
	0·050	4·35	3·49	3·10	2·87	2·71	2·60	2·51	2·45	2·39
	0·025	5·87	4·46	3·86	3·51	3·29	3·13	3·01	2·91	2·84
	0·010	8·10	5·85	4·94	4·43	4·10	3·87	3·70	3·56	3·46
	0·005	9·94	6·99	5·82	5·17	4·76	4·47	4·26	4·09	3·96
21	0·100	2·96	2·57	2·36	2·23	2·14	2·08	2·02	1·98	1·95
	0·050	4·32	3·47	3·07	2·84	2·68	2·57	2·49	2·42	2·37
	0·025	5·83	4·42	3·82	3·48	3·25	3·09	2·97	2·87	2·80
	0·010	8·02	5·78	4·87	4·37	4·04	3·81	3·64	3·51	3·40
	0·005	9·83	6·89	5·73	5·09	4·68	4·39	4·18	4·01	3·88
22	0·100	2·95	2·56	2·35	2·22	2·13	2·06	2·01	1·97	1·93
	0·050	4·30	3·44	3·05	2·82	2·66	2·55	2·46	2·40	2·34
	0·025	5·79	4·38	3·78	3·44	3·22	3·05	2·93	2·84	2·76
	0·010	7·95	5·72	4·82	4·31	3·99	3·76	3·59	3·45	3·35
	0·005	9·73	6·81	5·65	5·02	4·61	4·32	4·11	3·94	3·81
23	0·100	2·94	2·55	2·34	2·21	2·11	2·05	1·99	1·95	1·92
	0·050	4·28	3·42	3·03	2·80	2·64	2·53	2·44	2·37	2·32
	0·025	5·75	4·35	3·75	3·41	3·18	3·02	2·90	2·81	2·73
	0·010	7·88	5·66	4·76	4·26	3·94	3·71	3·54	3·41	3·30
	0·005	9·63	6·73	5·58	4·95	4·54	4·26	4·05	3·88	3·75
24	0·100	2·93	2·54	2·33	2·19	2·10	2·04	1·98	1·94	1·91
	0·050	4·26	3·40	3·01	2·78	2·62	2·51	2·42	2·36	2·30
	0·025	5·72	4·32	3·72	3·38	3·15	2·99	2·87	2·78	2·70
	0·010	7·82	5·61	4·72	4·22	3·90	3·67	3·50	3·36	3·26
	0·005	9·55	6·66	5·52	4·89	4·49	4·20	3·99	3·83	3·69
25	0·100	2·92	2·53	2·32	2·18	2·09	2·02	1·97	1·93	1·89
	0·050	4·24	3·39	2·99	2·76	2·60	2·49	2·40	2·34	2·28
	0·025	5·69	4·29	3·69	3·35	3·13	2·97	2·85	2·75	2·68
	0·010	7·77	5·57	4·68	4·18	3·85	3·63	3·46	3·32	3·22
	0·005	9·48	6·60	5·46	4·84	4·43	4·15	3·94	3·78	3·64
26	0·100	2·91	2·52	2·31	2·17	2·08	2·01	1·96	1·92	1·88
	0·050	4·23	3·37	2·98	2·74	2·59	2·47	2·39	2·32	2·27
	0·025	5·66	4·27	3·67	3·33	3·10	2·94	2·82	2·73	2·65
	0·010	7·72	5·53	4·64	4·14	3·82	3·59	3·42	3·29	3·18
	0·005	9·41	6·54	5·41	4·79	4·38	4·10	3·89	3·73	3·60
27	0·100	2·90	2·51	2·30	2·17	2·07	2·00	1·95	1·91	1·87
	0·050	4·21	3·35	2·96	2·73	2·57	2·46	2·37	2·31	2·25
	0·025	5·63	4·24	3·65	3·31	3·08	2·92	2·80	2·71	2·63
	0·010	7·68	5·49	4·60	4·11	3·78	3·56	3·39	3·26	3·15
	0·005	9·34	6·49	5·36	4·74	4·34	4·06	3·85	3·69	3·56

				Numerator df							
10	12	15	20	24	30	40	60	120	∞	P	df
1·96	1·91	1·86	1·81	1·79	1·76	1·73	1·70	1·67	1·63	0·100	19
2·38	2·31	2·23	2·16	2·11	2·07	2·03	1·98	1·93	1·88	0·050	
2·82	2·72	2·62	2·51	2·45	2·39	2·33	2·27	2·20	2·13	0·025	
3·43	3·30	3·15	3·00	2·92	2·84	2·76	2·67	2·58	2·49	0·010	
3·93	3·76	3·59	3·40	3·31	3·21	3·11	3·00	2·89	2·78	0·005	
1·94	1·89	1·84	1·79	1·77	1·74	1·71	1·68	1·64	1·61	0·100	20
2·35	2·28	2·20	2·12	2·08	2·04	1·99	1·95	1·90	1·84	0·050	
2·77	2·68	2·57	2·46	2·41	2·35	2·29	2·22	2·16	2·09	0·025	
3·37	3·23	3·09	2·94	2·86	2·78	2·69	2·61	2·52	2·42	0·010	
3·85	3·68	3·50	3·32	3·22	3·12	3·02	2·92	2·81	2·69	0·005	
1·92	1·87	1·83	1·78	1·75	1·72	1·69	1·66	1·62	1·59	0·100	21
2·32	2·25	2·18	2·10	2·05	2·01	1·96	1·92	1·87	1·81	0·050	
2·73	2·64	2·53	2·42	2·37	2·31	2·25	2·18	2·11	2·04	0·025	
3·31	3·17	3·03	2·88	2·80	2·72	2·64	2·55	2·46	2·36	0·010	
3·77	3·60	3·43	3·24	3·15	3·05	2·95	2·84	2·73	2·61	0·005	
1·90	1·86	1·81	1·76	1·73	1·70	1·67	1·64	1·60	1·57	0·100	22
2·30	2·23	2·15	2·07	2·03	1·98	1·94	1·89	1·84	1·78	0·050	
2·70	2·60	2·50	2·39	2·33	2·27	2·21	2·14	2·08	2·00	0·025	
3·26	3·12	2·98	2·83	2·75	2·67	2·58	2·50	2·40	2·31	0·010	
3·70	3·54	3·36	3·18	3·08	2·98	2·88	2·77	2·66	2·55	0·005	
1·89	1·84	1·80	1·74	1·72	1·69	1·66	1·62	1·59	1·55	0·100	23
2·27	2·20	2·13	2·05	2·01	1·96	1·91	1·86	1·81	1·76	0·050	
2·67	2·57	2·47	2·36	2·30	2·24	2·18	2·11	2·04	1·97	0·025	
3·21	3·07	2·93	2·78	2·70	2·62	2·54	2·45	2·35	2·26	0·010	
3·64	3·47	3·30	3·12	3·02	2·92	2·82	2·71	2·60	2·48	0·005	
1·88	1·83	1·78	1·73	1·70	1·67	1·64	1·61	1·57	1·53	0·100	24
2·25	2·18	2·11	2·03	1·98	1·94	1·89	1·84	1·79	1·73	0·050	
2·64	2·54	2·44	2·33	2·27	2·21	2·15	2·08	2·01	1·94	0·025	
3·17	3·03	2·89	2·74	2·66	2·58	2·49	2·40	2·31	2·21	0·010	
3·59	3·42	3·25	3·06	2·97	2·87	2·77	2·66	2·55	2·43	0·005	
1·87	1·82	1·77	1·72	1·69	1·66	1·63	1·59	1·56	1·52	0·100	25
2·24	2·16	2·09	2·01	1·96	1·92	1·87	1·82	1·77	1·71	0·050	
2·61	2·51	2·41	2·30	2·24	2·18	2·12	2·05	1·98	1·91	0·025	
3·13	2·99	2·85	2·70	2·62	2·54	2·45	2·36	2·27	2·17	0·010	
3·54	3·37	3·20	3·01	2·92	2·82	2·72	2·61	2·50	2·38	0·005	
1·86	1·81	1·76	1·71	1·68	1·65	1·61	1·58	1·54	1·50	0·100	26
2·22	2·15	2·07	1·99	1·95	1·90	1·85	1·80	1·75	1·69	0·050	
2·59	2·49	2·39	2·28	2·22	2·16	2·09	2·03	1·95	1·88	0·025	
3·09	2·96	2·81	2·66	2·58	2·50	2·42	2·33	2·23	2·13	0·010	
3·49	3·33	3·15	2·97	2·87	2·77	2·67	2·56	2·45	2·33	0·005	
1·85	1·80	1·75	1·70	1·67	1·64	1·60	1·57	1·53	1·49	0·100	27
2·20	2·13	2·06	1·97	1·93	1·88	1·84	1·79	1·73	1·67	0·050	
2·57	2·47	2·36	2·25	2·19	2·13	2·07	2·00	1·93	1·85	0·025	
3·06	2·93	2·78	2·63	2·55	2·47	2·38	2·29	2·20	2·10	0·010	
3·45	3·28	3·11	2·93	2·83	2·73	2·63	2·52	2·41	2·29	0·005	

Percentage points F_α of the F distribution (cont.)

Denominator df	Probability of a larger F	Numerator df								
		1	2	3	4	5	6	7	8	9
28	0·100	2·89	2·50	2·29	2·16	2·06	2·00	1·94	1·90	1·87
	0·050	4·20	3·34	2·95	2·71	2·56	2·45	2·36	2·29	2·24
	0·025	5·61	4·22	3·63	3·29	3·06	2·90	2·78	2·69	2·61
	0·010	7·64	5·45	4·57	4·07	3·75	3·53	3·36	3·23	3·12
	0·005	9·28	6·44	5·32	4·70	4·30	4·02	3·81	3·65	3·52
29	0·100	2·89	2·50	2·28	2·15	2·06	1·99	1·93	1·89	1·86
	0·050	4·18	3·33	2·93	2·70	2·55	2·43	2·35	2·28	2·22
	0·025	5·59	4·20	3·61	3·27	3·04	2·88	2·76	2·67	2·59
	0·010	7·60	5·42	4·54	4·04	3·73	3·50	3·33	3·20	3·09
	0·005	9·23	6·40	5·28	4·66	4·26	3·98	3·77	3·61	3·48
30	0·100	2·88	2·49	2·28	2·14	2·05	1·98	1·93	1·88	1·85
	0·050	4·17	3·32	2·92	2·69	2·53	2·42	2·33	2·27	2·21
	0·025	5·57	4·18	3·59	3·25	3·03	2·87	2·75	2·65	2·57
	0·010	7·56	5·39	4·51	4·02	3·70	3·47	3·30	3·17	3·07
	0·005	9·18	6·35	5·24	4·62	4·23	3·95	3·74	3·58	3·45
40	0·100	2·84	2·44	2·23	2·09	2·00	1·93	1·87	1·83	1·79
	0·050	4·08	3·23	2·84	2·61	2·45	2·34	2·25	2·18	2·12
	0·025	5·42	4·05	3·46	3·13	2·90	2·74	2·62	2·53	2·45
	0·010	7·31	5·18	4·31	3·83	3·51	3·29	3·12	2·99	2·89
	0·005	8·83	6·07	4·98	4·37	3·99	3·71	3·51	3·35	3·22
60	0·100	2·79	2·39	2·18	2·04	1·95	1·87	1·82	1·77	1·74
	0·050	4·00	3·15	2·76	2·53	2·37	2·25	2·17	2·10	2·04
	0·025	5·29	3·93	3·34	3·01	2·79	2·63	2·51	2·41	2·33
	0·010	7·08	4·98	4·13	3·65	3·34	3·12	2·95	2·82	2·72
	0·005	8·49	5·79	4·73	4·14	3·76	3·49	3·29	3·13	3·01
120	0·100	2·75	2·35	2·13	1·99	1·90	1·82	1·77	1·72	1·68
	0·050	3·92	3·07	2·68	2·45	2·29	2·17	2·09	2·02	1·96
	0·025	5·15	3·80	3·23	2·89	2·67	2·52	2·39	2·30	2·22
	0·010	6·85	4·79	3·95	3·48	3·17	2·96	2·79	2·66	2·56
	0·005	8·18	5·54	4·50	3·92	3·55	3·28	3·09	2·93	2·81
∞	0·100	2·71	2·30	2·08	1·94	1·85	1·77	1·72	1·67	1·63
	0·050	3·84	3·00	2·60	2·37	2·21	2·10	2·01	1·94	1·88
	0·025	5·02	3·69	3·12	2·79	2·57	2·41	2·29	2·19	2·11
	0·010	6·63	4·61	3·78	3·32	3·02	2·80	2·64	2·51	2·41
	0·005	7·88	5·30	4·28	3·72	3·35	3·09	2·90	2·74	2·62

Table IV is taken from Merrington and Thompson (1943) ,'Tables of percentage points of the inverted beta (*F*) distribution', *Biometrika*, **33** and from Table 18 of *Biometrika Tables for Statisticians*, Vol. 1, Cambridge University Press, 1954, edited by E. S. Pearson and H. O. Hartley. Reproduced with permission of the authors, editors and *Biometrika* trustees.

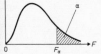

				Numerator df							
10	12	15	20	24	30	40	60	120	∞	P	df
1·84	1·79	1·74	1·69	1·66	1·63	1·59	1·56	1·52	1·48	0·100	28
2·19	2·12	2·04	1·96	1·91	1·87	1·82	1·77	1·71	1·65	0·050	
2·55	2·45	2·34	2·23	2·17	2·11	2·05	1·98	1·91	1·83	0·025	
3·03	2·90	2·75	2·60	2·52	2·44	2·35	2·26	2·17	2·06	0·010	
3·41	3·25	3·07	2·89	2·79	2·69	2·59	2·48	2·37	2·25	0·005	
1·83	1·78	1·73	1·68	1·65	1·62	1·58	1·55	1·51	1·47	0·100	29
2·18	2·10	2·03	1·94	1·90	1·85	1·81	1·75	1·70	1·64	0·050	
2·53	2·43	2·32	2·21	2·15	2·09	2·03	1·96	1·89	1·81	0·025	
3·00	2·87	2·73	2·57	2·49	2·41	2·33	2·23	2·14	2·03	0·010	
3·38	3·21	3·04	2·86	2·76	2·66	2·56	2·45	2·33	2·21	0·005	
1·82	1·77	1·72	1·67	1·64	1·61	1·57	1·54	1·50	1·46	0·100	30
2·16	2·09	2·01	1·93	1·89	1·84	1·79	1·74	1·68	1·62	0·050	
2·51	2·41	2·31	2·20	2·14	2·07	2·01	1·94	1·87	1·79	0·025	
2·98	2·84	2·70	2·55	2·47	2·39	2·30	2·21	2·11	2·01	0·010	
3·34	3·18	3·01	2·82	2·73	2·63	2·52	2·42	2·30	2·18	0·005	
1·76	1·71	1·66	1·61	1·57	1·54	1·51	1·47	1·42	1·38	0·100	40
2·08	2·00	1·92	1·84	1·79	1·74	1·69	1·64	1·58	1·51	0·050	
2·39	2·29	2·18	2·07	2·01	1·94	1·88	1·80	1·72	1·64	0·025	
2·80	2·66	2·52	2·37	2·29	2·20	2·11	2·02	1·92	1·80	0·010	
3·12	2·95	2·78	2·60	2·50	2·40	2·30	2·18	2·06	1·93	0·005	
1·71	1·66	1·60	1·54	1·51	1·48	1·44	1·40	1·35	1·29	0·100	60
1·99	1·92	1·84	1·75	1·70	1·65	1·59	1·53	1·47	1·39	0·050	
2·27	2·17	2·06	1·94	1·88	1·82	1·74	1·67	1·58	1·48	0·025	
2·63	2·50	2·35	2·20	2·12	2·03	1·94	1·84	1·73	1·60	0·010	
2·90	2·74	2·57	2·39	2·29	2·19	2·08	1·96	1·83	1·69	0·005	
1·65	1·60	1·55	1·48	1·45	1·41	1·37	1·32	1·26	1·19	0·100	120
1·91	1·83	1·75	1·66	1·61	1·55	1·50	1·43	1·35	1·25	0·050	
2·16	2·05	1·94	1·82	1·76	1·69	1·61	1·53	1·43	1·31	0·025	
2·47	2·34	2·19	2·03	1·95	1·86	1·76	1·66	1·53	1·38	0·010	
2·71	2·54	2·37	2·19	2·09	1·98	1·87	1·75	1·61	1·43	0·005	
1·60	1·55	1·49	1·42	1·38	1·34	1·30	1·24	1·17	1·00	0·100	∞
1·83	1·75	1·67	1·57	1·52	1·46	1·39	1·32	1·22	1·00	0·050	
2·05	1·94	1·83	1·71	1·64	1·57	1·48	1·39	1·27	1·00	0·025	
2·32	2·18	2·04	1·88	1·79	1·70	1·59	1·47	1·32	1·00	0·010	
2·52	2·36	2·19	2·00	1·90	1·79	1·67	1·53	1·36	1·00	0·005	

Index